RUCK UP
OR SHUT UP

A COMPREHENSIVE GUIDE TO
SPECIAL FORCES ASSESSMENT AND SELECTION

Copyright © 2023 by David Walton
All Rights Reserved
No portion of this book may be reproduced in any form without written permission from the publisher or author, except as permitted by U.S. copyright law.

Cover Design: Brad B
Photo credits K Kassens, USAJFKSWCS

ISBN: 979-8-9881260-0-3

DISCLAIMER: SFAS is hard. I make no guarantees about the results of the information applied in this book. I share educational and informational resources that are intended to help you succeed in SFAS. You need to know that your ultimate success or failure will be the result of your own efforts, your particular situation, and innumerable other circumstances beyond my knowledge and control. Your adequate preparation will require immense physical effort. You should consult with a physician to determine your suitability prior to any training.

This book is not endorsed by the Department of Defense, the United States Army Special Operations Command, or the United States Army John F Kennedy Special Warfare Center and School…but it should be.

For my family.

My wife, Kristen, who served right beside me.
My kids, Jack and Lauren, who
didn't always have dad home and they turned out perfectly.
Thanks to Kristen.

They are my *Why*.

The 3 Rules

Always look cool.

Never get lost.

If you get lost, look cool.

The SOF Truths

Humans Are More Important Than Hardware

Quality is Better Than Quantity

Special Operations Forces Cannot Be Mass Produced

Competent Special Operations Forces Cannot Be Created After Emergencies Occur

Most Special Operations Require Non-SOF Support

SOF Imperatives

Understand the operational environment

Recognize political implications

Facilitate interagency activities

Engage the threat discriminately

Consider long-term effects

Ensure legitimacy and credibility of Special Operations

Anticipate and control psychological effects

Apply capabilities indirectly

Develop multiple options

Ensure long-term sustainment

Provide sufficient intelligence

Balance security and synchronization

Prologue 13

Chapter 1: Introduction 19

Chapter 1.5 - A Word About "We" 36

Chapter 2 – Special Forces – The Culture 38

Chapter 3 – The Upside-Down World 80

Chapter 4 - How to Prepare for SFAS – Gate Week 136

Chapter 5 - How to Prepare for SFAS – Land Nav Week 206

Chapter 6 - How to Prepare for SFAS – Team Week 228

Chapter 7: Can't Tell Me Nothin', Can't Show Me Nothin' 253

Chapter 8: The Lame, the Blind, the Mute, and the Crippled . 262

Chapter 9: The Young and the Restless 275

Chapter 10: Cadre 300

Chapter 11: The Standard is the Standard 313

The Last Chapter: The Why 326

A Note About Terms, Definitions, and Language

This book isn't a peer-reviewed academic journal submission. Neither is it a sanitized account for mass-media consumption. You won't read censored and curated feel-good tales. This book is a prescriptive account of SFAS and a descriptive account of the culture, legend, and lore that surround SFAS. There are scatological anecdotes, ample cursing, and generally vulgar descriptions. There are stories of agony and misery. SFAS is hard and being a Green Beret is a difficult lifestyle full of hardship. The language in this book reflects this reality. If you are reading this book, you are likely exploring a future in Special Forces. We say fuck a lot, get used to it.

I would also note the repeated use of the term *guy* or *guys*. I understand that this is a gendered term, and that sort of thing usually gets frowned upon. I don't care. I use it here as an inclusive term. That is the way the military works and this is a work of military significance. I also happen to be one of the most informed researchers on women in Special Forces. I helped form the charter for the USASOC Cultural Support Teams, I was a featured speaker at the US Institute for Peace on Women, Peace, and Security, I have consulted on multiple gender integration research projects, and I conducted the most comprehensive study of female Candidate integration into SFAS to date. You can read of my work in this field at War on The Rocks. No one is more qualified to speak on this topic than I am, but this is not the thesis of this book. So, *guys* it is.

You may also note that I make no claims about this book doubling your chances of getting selected or any such nonsense. Some books and prep programs do make this claim. My response would be, "Prove it". First, doubling compared to what? Not reading the book? Not doing the program? How on earth are you measuring this? Do you track every person that buys your product and compare them to the general SFAS population? Are you honestly making the claim that 72% of your customers or clients get selected? What an obscene marketing tool. What an outrageous assertion. If they are willing to make that claim on something so contemptibly unbelievable, then what else are they willing to lie about?

I also assume that you are capable of researching what an ODA is, or what SWCS is, or how the pipeline generally works. I don't want to spend effort retelling tales that are best told elsewhere. The internet exists and we live in a rich information environment. There are so many other great books that can do this job. My specialty is SFAS, and I want to talk about this specifically. Only one book can tell this story of SFAS. This is that book.

Good men will not just sit and watch other men suffer.

Prologue

Are you elite?

Before you answer that question, you need to understand what that word means for Green Berets. You need to decide if you are willing to make the sacrifices required to be elite. The sacrifices are immense. It's delayed gratification. You won't get to enjoy every holiday at home. You won't get to partake in every beer and stay out late all the time. You must be a professional tactical athlete and all that this entails. Grinding workouts and endless effort. Special Forces Assessment and Selection (SFAS) is just the beginning. Good enough will never be good enough. Being elite means that you never say, "Can I?" You always say, "I can." And then you do it.

Not everyone who reads this book will go to SFAS. There is still much to enjoy in the following chapters. There are great stories and lessons to be learned for just about anyone. For those who are already in the pipeline for Selection, these pages are a manual for what you can expect at Camp Mackall. For those on the path but not yet committed, you will likely find words of motivation and confirmation that will help you to commit. For those that are searching for something more, know that you have found it. If you discovered that your current station in life doesn't fulfill you, this book is a map to self-actualization. If you are staring down the barrel of a not-quite mid-life crisis and wondering if there is something that you missed, you can find it here. It is never too late to start the journey to being elite.

But this book isn't about me, it's about you. That realization is an immense burden. If you bought this book because you wanted to hear about someone else's journey, then you bought the wrong book. This is about *your* journey. What you need to be willing to do. What the challenges are that lay ahead for you. I have already made my journey. My path has been walked. I have already lived amongst the elite. This is about you. If you want to be elite, then you need to map your course now. You need to figure out how to get your Green Beret.

The Green Berets are a Brotherhood. Born in the shadowy world of the British Special Operations Executive and the US Office of Strategic Services of World War 2, the Green Beret is a capability like no other. As a Green Beret you will learn to shoot, move, communicate, and medicate at a level that you can hardly imagine. You will be lethal. But you will also learn to read a room and build the instinct to walk into any situation and figure out who the real decision makers are. You will learn to be dangerous and how to manage risk. A Warrior-Diplomat.

Kennedy And Yarborough, 1961, photo credit US Army

You will do all this better than anyone else. That is elite. President John F. Kennedy would no doubt be proud of the modern Green Beret. The man, the mission, and the hat. Kennedy is rightfully seen as a patron saint of sorts for the Green Beret. In October 1961, Kennedy visited Fort Bragg where he interacted with Green Berets who were wearing, at the time, the unauthorized headgear. He returned to the White House and penned his now famous Presidential Memorandum where he officially awarded the beret to the Regiment. He called the Green Beret "a symbol of excellence, a badge of courage, a mark of distinction in the fight for freedom." This mark of distinction is justly seen as elite. Courage is still a hallmark. The fight for freedom continues.

Why would you deliberately choose this path of hardship? If you decide to be a Green Beret, you will suffer. It is just part of the package. Your body will bear the brunt of the hardships, but your mind will suffer too. What would make you do this? When you commit to this endeavor, half-measures will not be enough. Total commitment is what it takes. Think on this deeply, because it is important. What is your *Why*?

Maybe you remember when you were a kid. You remember when you went swimming and you would practice holding your breath and then slipping out of the water as quietly as possible. You would do it over and over again until you were almost silent. Maybe you had family that served. You recall that sense of awe that you felt when you saw the uniform. Maybe you remember the sense of accomplishment when you built that cool fort on the empty lot in your neighborhood. You staked your claim to a small part of the world, and it was your kingdom. Maybe you remember the thrill of discovery when you learned how to ride a bike or throw a ball or sharpen a stick. You remember that you are capable of great things. You still are.

You grew up and so did your goals. Your *Why* has grown. You want something more. You know that you are capable of great things, but there is little greatness in your circle. You joined the Army because you wanted to test yourself, but all you get is open-book multiple choice exams. There is no challenge. You pursue excellence because you know that a life of purpose is the only thing that matters. I want you to read this book and think about your *Why*. When you read the final chapter, maybe it will be a little clearer for you. Maybe this book will teach you some things and maybe it will just help you teach some things to yourself. You are beginning a journey towards a great unknown and you have much to learn. It is a worthy endeavor.

If you want to be a Green Beret, then your pursuit of excellence begins at Special Forces Assessment and Selection.

Are you ready to be elite?

Chapter 1: Introduction

Why did I write this book?
Two reasons – misinformation and duty.

Special Forces Assessment and Selection (SFAS or simply Selection) remains a bit of an enigma. This is deliberate. We don't want everyone to know everything. It should be a little bit mysterious. Selection is designed to be a high barrier to entry into the Special Forces Regiment, so it makes good sense to not tell everyone exactly what to expect. But we ('we' being the SF Regiment, the Special Warfare Center and School, the powers that be, etc.) should be diligent in managing the messaging about Selection. We are not. We seem content to let the narrative manage itself. This is a bad idea.

Allowing the narrative around SFAS to develop organically has proven to be a disaster. There are a couple of reasons for this. This grassroots messaging campaign is run by the least informed people possible; the International Barracks Lawyers Guild, SFAS Non-Select Chapter. We are a victim of our own success in this regard. Selection is incredibly effective. We manage to successfully screen thousands of high-quality Candidates every year. This intensive, rigorous, and comprehensive process produces some by-product. Some waste material if you will. Historically, SFAS selects around 36% of Candidates who attend (United States Special Operations Command, 2018). So, 64% of the Candidates that attend don't get selected. Nearly two-thirds of the people exposed to this unique training environment leave with a likely bad-taste in their mouth. What happens to them? They go back to their units, *right back into our prime recruiting source*, and start talking.

    And what do these disgruntled Barracks Lawyers have to say? Whatever they want, because we haven't produced a cohesive messaging campaign to establish the true story. What do we expect them to say? That they tried their hardest and their very best wasn't good enough? Do we expect them to confess that it is their fault that they didn't make it? That they were treated entirely fairly, and the standards were just too high? Do we really believe that human nature would allow this? Or will they say things like "I don't know, man. They're just all fucked up out there at Camp Mackall." I think we have all heard the stories. I certainly have. I have had multiple unsuccessful Candidates earnestly tell me that they didn't get selected because the Cadre didn't like their tattoos. Or that the Officers have higher standards than the Enlisted. Or that the National Guard is hurting for bodies so they're just letting everyone in. Or that they're not combat arms, or they're a minority, or they're not a Ranger, or…or…or. You get the picture.

Can you imagine that? Can you imagine the perverted sense of standards that would allow a Cadre to just declare that he didn't want that guy to get through SFAS because of his ink? What would the reaction of fellow Cadre or leadership be if we just declared that only combat arms or only Rangers were allowed? Less than half of the Cadre have Ranger tabs themselves, and it is purely insulting to think that we would allow bigotry to influence our decisions. I simply cannot fathom why any reasonable person would accept the narrative that SFAS is so compromised that this sort of dereliction of duty occurs.

But that's the narrative. The least successful and least informed amongst us, the non-selects, are driving the messaging about SFAS. I spent two years conducting an exhaustive study of SFAS and in that process I interviewed hundreds and hundreds of Candidates. Every single Candidate that I interviewed, without exception, shared that they expected and were fully prepared to see strong biases from the Cadre. They had heard the narrative that the *Guild of Idiots* had carefully crafted prior to arriving at Camp

Mackall and were ready for the inevitable shit show. But not one, not a single successful Candidate, could cite a single instance of unfair treatment, bias, favoritism, or prejudice. Everyone expected it, but no one experienced it. Imagine that. That is a powerful message. But that's just one-third of the population.

    The other two-thirds were a different story. They were mostly confused. In most cases they knew why they didn't get selected. Maybe not the precise standard that they didn't meet, but they knew why in general terms. For many it doesn't matter. This is a crisis for this Soldier. He was working towards a goal, a dream even, and he didn't make it. He is going to have a natural, and frankly predictable, response. He is going to blame something or someone else. I personally witnessed innumerable Candidates that had just hours prior quit Selection, and then say they "got screwed." A Candidate would walk up to a Cadre and say, "Sergeant, I Voluntarily Withdraw," out-loud, in front of God and Country, and then saunter back to the barracks and bold-face claim to his peers that he got screwed. The drop-out barracks are a den of lies and deception. They are the norm.

The drop out barracks are the primordial birthplace of the Barracks Lawyers Guild[1]. It is here where the narrative begins, unchecked. Rumors start, biases are confirmed, and narratives are shaped. Candidates rally around each other to console and appease. It starts with, "I don't know why," and soon migrates to something darker. Before long, it is vitriol and outright fabrications. These spokespeople return to their units, directly into our target recruiting population, and tell their lies.

Because there is no counter-narrative, no facts, this story becomes the truth. Who is going to challenge this account? The potential recruits back in the operational force rely on word of mouth to understand SFAS. The aggrieved becomes the expert and his falsehoods become the reality. You might be surprised to learn how many Candidates claimed they were a medical drop. That is an honorable defeat, after all. You wanted to continue but your body just broke, and the Cadre made a medical decision to prevent you from doing irreparable damage to yourself. You can leave with your head held high. You probably would not be surprised to learn that the actual medical drop

---

[1] There are plans to build separate barracks for Candidate drop-outs, but they currently house in temporary deployable shelters away from active Candidates

rates are incredibly low. Less than 5% of Candidates are medically dropped from SFAS (USAJFKSWCS, SFAS Daily Summary, 08-22, 07-22, 02-22).

Social media has made this even more challenging. A lie can be half-way around the world before the truth ever puts his ruck on. Unfortunately, we seem to be content leaving our rucks untouched. I am all for being the Quiet Professional, but we are entirely silent. You can anonymously make a wild claim about Selection and the only evidence required is, "because I said so." But something interesting is happening. In the midst of my years long research project, the Cadre started to properly out-counsel dropped Candidates. Candidates were told why they were dropped. A formal one-on-one session with a senior Cadre member. A dropped Candidate may not be told the specific standard that he failed to meet, but he was told a reason. Almost instantly the information environment in the drop-out barracks changed. It was no longer feasible to be so blatant in the self-deceit, let alone the open lies. He knew why he didn't make it, and everyone else knew that he knew. And it certainly was not because of his tattoos. Not every class gets the same level of out-brief, but it is slowly changing.

Even better, some Candidates reported incredibly positive out-brief sessions. Cadre actually thanked Candidates for attending. Shook their hands, looked them in the eye, and thanked them. Why shouldn't they? We invited them to come to SFAS. We asked them to be here. They volunteered, but we screened their packets and told them that we wanted to take a better look at them. We recruited them. We sent them orders to come to Camp Mackall. This is an honorable exchange with noble intentions. If the Candidate didn't lie, cheat, or steal then we should uphold that honor. They didn't make it. So what? Most don't. There is no dishonor in this. Wouldn't it be acceptable to have them come back and try again? Wouldn't it be best to have them return to the operational force with a positive message. A message that SFAS is hard, but fair. You will be tested, but you'll be treated with decency and respect. The truth.

We should combine this grassroots message with an official public information campaign that highlights the realities of SFAS, without giving away the standards. This is not an insurmountable task. But instead, we simply don't engage at all. We just use *hope* as a method. We *hope* that the right message will be sent. We *hope* that the right guys will come. We *hope* that we can build and sustain the force. How is this working out for us so far? Are we meeting our recruiting goals? Are we satisfied with the current information environment? Are we pleased that most of the force thinks we are arrogant assholes with an agenda and no standards?

I'm not satisfied. So, that's one reason why I wrote this book.

The other reason is duty. I acknowledge that this is a loaded statement. In many circles it is considered vulgar to say that you love America. Someone who seeks to serve is often seen as a rube. It has somehow become awkward to admit that you are elite. Meritocracy is seen as bigotry. But I find this attitude disgraceful. I love America and all its blessings. I am deeply humbled and profoundly honored to be a Green Beret. The Regiment has given me far more than I could ever give it. It is not an easy life, but it is a worthy one. A life of purpose. So, I say this with complete sincerity.

You have a duty.

If you are one of the few people who are capable of earning the Green Beret, then you have a duty to do so. There are few who can. The Pentagon reports that 77% of Americans between 17-24 aren't even eligible to serve because they are too fat, too dumb, or too lawless (Defense, 2020). They use nicer phrasing, but that is the reality. Of the less than 1% of Americans who do serve, even fewer are eligible to apply for Special Forces (Mission: Readiness, 2009). There simply are not many people who even meet the bare minimum requirements. And then there is Selection. SFAS is a massive barrier to entry. It is nearly insurmountable.

Selection is designed to be, well…selective. We measure everything. We poke and prod. We challenge and record. We test and quantify. We judge. We are harsh critics with a keen eye, and we make you earn it. We have standards. Immovable standards. We haven't really changed our standards, ever. If you successfully emerge from the Pinelands of Camp Mackall, you will know that you have done something special. Something very few people can do. So, if you are one of the few who can, then you have a duty to try. I want you to know that you will be given a fair shot. We need you, and I suspect that you need us. So, I am writing this book to tell you all of this in plain language.

Bona Fides

Who am I? Why do I get to represent SFAS on these pages? Who appointed me to be the spokesman for the Green Beret?

Well, I am a Green Beret. I joined the Army in 1991 and after commissioning and a tour as a Cavalry officer I attended SFAS in 1997. I served in the 7th Special Forces Group (Airborne) and later in the Joint Special Operations Command. I retired out of the United States Army John F. Kennedy Special Warfare Center and School (USAJFKSWCS or SWCS) in 2013 as a Lieutenant Colonel. I did all the cool guy schools and all the deployments. Of my two decades on active duty, I spent most of them away from my home and my family. I have an impressive shadow box of badges, tabs, and awards and fond memories of a worthy career.

While serving at SWCS I had oversight over SFAS. I use the word *oversight* specifically because it means something. During my time at SWCS I had a large portfolio. I was entrusted with leading SWCS's Education, Regional Studies and Culture, and Human Dynamics efforts. I earned the trust of senior leadership and SFAS was a bit of a homeless entity, so it came under my oversight. This was during a critical phase in its development. All of SWCS, but especially the Special Forces pipeline, was subjected to a significant effort to realign the generating force with what the operating force of the future required. A strategic reckoning.

So SWCS was digging into every nook and cranny of what we did, how we did it, and most importantly *why* we did it. We realized that we were sprinting so fast to keep up with the wars in Iraq and Afghanistan that we had allowed many unnecessary things to sneak into the pipeline. SFAS was a critical component in this discussion, even though it remained largely unchanged since its creation. I was fortunate that the SFAS Commander and Sergeant Major were incredibly talented, and I quickly realized my best play was to listen to them, get them the resources they needed, and otherwise stay out of the way. But while doing this, I learned. While the SFAS Cadre asked and answered the tough questions about what we did, what it measured, and why we did it, I learned. I got to weigh in on key decisions and shape some elements, but the effort was much bigger than me. What emerged is the current SFAS, largely unchanged, but now quantified and understood. We always knew that SFAS worked, but now we knew why.

And then I retired. I went on to finish my Doctorate and my dissertation research was centered on SFAS. I put my degree to good use. I worked for the National Defense University at SWCS and used my newly honed research and analytical skills to continue to study SFAS from the inside. I have conducted hundreds of hours of SFAS observation, hundreds of interviews of Candidates, Cadre, and Commanders, and I have seen every bit of the data that we collect. I know all the methodology, all the factor analysis, and all the standards. My notes include copies of the training schedule, SFAS atmospherics, data management practices, and assessment metrics. I talk to Cadre and Candidates almost every day. I know what goes on at Camp Mackall and more importantly I understand what it means. I would not say that I am *the* expert in SFAS, but I am certainly *an* expert in SFAS.

    I also recognize that I signed a Non-Disclosure Agreement (NDA) specifying that I would not reveal any secrets. So, while I will describe a great deal about how you can prepare for SFAS, I will not tell you any secrets about SFAS. *All the rules matter all the time* and I intend to honor my obligations. But there is already a great deal published about Selection. There is a plethora of junk out there. Well-meaning Podcasts and cringeworthy motivational videos are common. Clickbait articles, conjecture, opinion, and outright disinformation abounds.

Internet forums are packed with guys that know a guy or heard from another guy. Wisdom of the ancients. That stuff is typically useless. Most of the useful data is buried in academic articles. The academic material is a bit hard to translate and it needs a lot of context to really be useful. But it is accurate. It is endorsed by SWCS. And it is not covered by the NDA.

So, I intend to use my decades of education and experience, my unique placement and access, and my research and writing skills to help battle some of the misinformation about SFAS. The truth is much more compelling than the narrative. And it might help you to do your duty. It might just help you get selected.

Who Is This Book For?

This book is for the Special Forces Regiment, present and future. The Brotherhood should know that despite the tale that SWCS has somehow lost control over the pipeline, that standards have eroded, and that the Green Beret is but a shell of what it once meant, none of that is true. Special Forces Assessment and Selection is as hard as it has ever been. I have seen the evidence, all the evidence, and I can attest that the standards have not changed. In fact, I would assess that Selection is harder than ever. Read on if you want to know why. I have walked countless miles and watched countless events and talked to countless people. I have two decades of dedicated study of SFAS. I have investigated every claim that I could and looked in every corner of Camp Mackall. **My Brothers, I can affirm that the Green Beret remains a symbol of excellence, a badge of courage, and a mark of distinction in the fight for freedom**.

This book is also for the hopeful few who will dare to try. There is a rare breed of American who wants to serve. They feel the burden of duty. They know that a life of grand purpose is possible. They may not know what a Green Beret really does or precisely what it takes to earn the coveted title, but they feel the intense need to test their mettle. They want to be part of the Brotherhood. I have included in this book a section just for the young aspiring teenager who is looking for something more. I would encourage everyone to carefully read this section as well, as it will tell you a great deal about what type of society we must endeavor to foster if we want to sustain this great Nation. This assessment is not an opinion, it is based on the empirical data of what we look for in Special Forces Candidates.

Most importantly this book is for those who are ready to make the commitment. You have done the research, you have talked to your family, and you are ready to begin your journey. You know you want to test yourself, but you don't know where to start. The good news is that if you are reading this book then you are headed in the right direction. Keep reading. Then ruck up, we have work to do.

No Shit, There I Was...The War Story and the Power of Storytelling

The war story is a staple of military life. The average legionnaire's existence is basically *hurry up and wait*. Everything is an emergency and requires immediate action, followed by standing around doing absolutely nothing. Sheer panic and terror, punctuated by utter boredom and idle hands. I think that this is a Universal Truth for soldiers everywhere.

This lifestyle provides for ample opportunity to tell stories to fill the time. We call them *war stories*, but they often don't involve actual war. The emphasis is on the story. A good storyteller is a valuable asset to a unit, providing much needed entertainment during dire times. A middling soldier who can spin a good yarn enjoys an elevated status amongst his peers. A good soldier who can tell good stories is worth his weight in gold.

There are some rules to war stories. They should begin with the cue to learning...something like "That reminds me of this one time...," or "True Story, listen to this...," or "My buddy told me about a guy who...." But the best attention getter is "No shit, there I was...." This is sometimes expanded to "No shit, there I was...knee deep in spent brass and hand grenade pins...." When you hear this, you can be assured of less *war* and more *story*. I prefer the shorter version.

Another rule is that it must be at least partially true, but it is more important that they convey a sentiment. The nature of war stories, like the nature of all oral history, is that they get passed down and spoken to many ears. This is especially true of war stories, and in the retelling, they sometimes get a little distorted. They are grounded in fact, but sometimes not by much. It can be difficult to separate the wheat from the chaff because so much of what a soldier is subjected to is just absurd. I think the official rule is that they must be at least 10% true at least 10% of the time.

I am going to use this time-honored storytelling tradition in this book. Some of them are mine and some of them have become mine. This is just the way that war stories work. War stories can be a helpful tool by punctuating a point or emphasizing a theme. They can also be wildly entertaining. My goal for this book is to help as many people as I can.

I promise not to take too much liberty with the rules.

Chapter 1.5 - A Word About "We"

You may note that I flip between using I, we, and they (Cadre, SWCS, the Army, etc) throughout this book. Sometimes this is purposeful and sometimes this is just the vagaries of language and the challenges of communication. I spend some time in the next chapter talking about the importance of clarity in communication, so let me be absolutely clear. I am not Cadre. I have never been Cadre. I have spent significant time at SFAS and walked many events and conducted countless hours of observation. I played a small role in shaping SFAS during my time at SWCS when SFAS was part of my leadership portfolio, but I was not Cadre. I also don't have any of the limitations of Cadre. I didn't have to contend with the nuts and bolts of administering SFAS, so I have a unique perspective of the totality of the enterprise. This unique perspective is helpful in that I understand more about the *why* because I don't have to worry about the *how*. I can make a much more holistic assessment of PT test administration because I wasn't restricted to just counting pushups. So, this book isn't about how to grade a PT test, it's about how PT tests are graded, so to speak.

I also recognize that being a Green Beret is not always a casual endeavor. For some, they don't feel any particular connection to the Regiment after they leave service. It was just something that they once did. For others, being a Green Beret is their entire identity. They build businesses on the connection, they plaster their car with stickers, or they just enjoy the inevitable accolades and privileges that accompany the title. Different strokes for different folks. I think I fall somewhere in the middle. I appreciate that I have a full life away from this book and the beret. But I also feel a profound sense of duty to continue to shape the Regiment. That is one of the reasons why I wrote the book; to convince the next generation of Green Berets that they should commit to this lifestyle and if they do, they will be treated with dignity and respect. There are no red carpets at Camp Mackall, but no one will pull the carpet out from under you either. So, in many ways I, and many others like me, are still very much a part of the "we". It is my Regiment, too. And I have a duty to attend to the Regiment's business. I am retired, but I am not expired.

Chapter 2 – Special Forces – The Culture

The history, organization, and mission of Special Forces is well documented and readily available elsewhere. It would be incredibly low effort to reproduce it here and then count it as part of a contribution to the body of knowledge. If you want line and block organizational charts, lists of mission sets, descriptions of the various roles and responsibilities, or the boundaries of the geographic areas of responsibility then you will have to look elsewhere. Even if you went to the worst sources possible and learned completely inaccurate stuff, it wouldn't impact your chances of getting selected. The Sandman is not any lighter if you know which Special Forces Group historically deploys to Africa or which military occupational specialty manages communications. You could know everything that there is to know, but you still need to put in the work to become part of that knowledge not just the recipient of it. We will focus on the topics that rarely get covered yet help define so much about the culture, lifestyle, and reality of Special Forces.

The Green Beret

First let's address the Green Beret. Berets, writ large, are horrible. They don't shade the eyes from the sun. They are hot. They stink when they get wet. They require both hands to properly don. Unless you spend significant effort shaving and shaping the beret in preparation for wear, they look bad. If you put it on wrong, you just look silly. We have all seen the poor sap who didn't follow the rules and his beret looks like a chef's hat or a mushroom cap. In my day, the beret was purely ceremonial. You only wore it a few times a year for special occasions. We often went six months or more with only a patrol cap. This has changed of late. Starting in the mid-2000s leaders started to push for its wear more often. It slowly became the standard for garrison wear, as it remains today.

Maybe this was a reaction to the 2001 Army Chief of Staff Eric Shinseki decision to replace the 75th Ranger Regiment's vaunted black beret with a tan one, and then 'award' that storied piece of black felt to the rest of the Army (Murphy, 2019). Elite units wear berets, so if a unit wears a beret, then it must be elite, right? At the time, this move was earth-shatteringly controversial. The Ranger Regiment takes great pride in their appearance and unit history. The black beret was a significant part of their identity. A newly arrived Ranger Private was given specific instructions on how the black beret was to be prepared and worn. His team leader was charged with the direct supervision of the black beret's preparation. When the black beret was finally prepared to specification and the Private was allowed to wear it in public, the black beret always looked right. The correct angle perched high off the back of the head, the front draped just so over the right eye, and the form shaped closely to the head. I have never seen a Ranger, even to this day, with a bad beret.

But at the time, this move was a significant emotional event for many. The backlash was extreme, immediate, and ultimately ineffective. The Rangers lost their black beret. So maybe senior Special Forces leaders wanted to reinforce the Green Beret as a trademark and a protected entity of Special Forces. Use it or lose it, in a way. A practical stance, one might conclude. President Kennedy memorialized the Green Beret's distinction, and it is ours, forever. But in making it an everyday accessory it has almost become a status symbol rather than an ideal to be upheld. The Green Beret is a representation of excellence; a model for a life well lived. It represents something greater than just a hat.

"A SYMBOL OF EXCELLENCE, A BADGE OF COURAGE, A MARK OF DISTINCTION IN THE FIGHT FOR FREEDOM."

U.S. ARMY PHOTO BY SPC. DYLAN SMEIGH

I recall when I went to SFAS that during one of the endless late-night sessions where a lone Cadre was charged with keeping us awake for the sake of keeping us awake, we were moved into a large classroom. We were assigned the task of drawing on our piece of paper a picture of what we thought being Special Forces was all about, the beret or a patrol cap. We had two hours to craft our masterpiece sketch of the headgear that most represented our purpose at SFAS. When the Cadre was giving us his instructions it became quite clear that the expectation was absolute; you will draw a patrol cap. The patrol cap represented hard work. Grinding days in the field, demanding austere deployments, and selfless service to the Special Forces Regiment. If you were to make the fatal error of drawing a beret then you were sending a clear message that you thought being Special Forces was all about the pomp and circumstance, the glory, and the accolades. You were at SFAS for the wrong reasons.

Nobody drew a beret, and the message was seared into my psyche. You are here to serve the Regiment. Whatever benefits you may gain from this service are entirely tangential to the mission. This attitude was reinforced throughout my training and assignments. You serve at the pleasure of the SF Regiment, and you are not authorized to spend any of the currency of credibility that our reputation provides us. You can add to that credibility bank account, but you may not withdraw from it. I think that we may have lost that notion in recent years. Earning the Green Beret is seen by some as a steppingstone to something else. A means to another end.

So, I would give the aspiring Candidate the following advice. Make certain that you are on this journey for the right reasons. There are dark days ahead for you. SFAS can be an existential crisis when the weight of an apparatus is relentlessly pressing you into the sand. The gratification of social media notifications and likes will not fuel you adequately. Team life can be an incredible burden. You will make the deepest connections and the strongest bonds with your Brothers. You will willingly give your life for them and them you. And it will happen. In the emotional aftermath of this trauma, you will need the sort of resiliency that vapid motives will not provide. You will sacrifice your body in this journey. Physical disability is the inevitable product of jumping out of planes, blowing stuff up, and a lifetime of hard routine. Make certain that your motives are proper. Patrol cap, not beret.

SF vs SOF

This may sound petty, and in some ways, it is. But words have meaning, and we should endeavor to be as precise as possible in our language. Green Berets are Special Forces. Special Forces are Green Berets. Capital S, capital F. This is different from special forces, all lowercase. Lowercase special forces is a generic term to describe forces that do special things. Upper case Special Forces are people. Special people that do special things. A rare breed. Lower case special forces is a description. When you use the terms interchangeably it diminishes what we do. It is not accurate. This is exacerbated when one uses the term Special Operations Forces. SF vs SOF. Green Berets are Special Operations Forces, but only a very few Special Operations Forces are Green Berets. Most people don't know this or don't care to know it. The terms get interchanged, and the meanings get diminished.

    This is not akin to the 'is a hot dog a sandwich' debate (it is not). It is minor, but it is not an inconsequential battle. This is about protecting our credibility. We have struggles enough protecting the brand image when we just account for actual Special Forces. But when every enabler in the formation starts to claim, "Yeah, I'm in Group" and they use the convoluted SF vs SOF vs lower case explanation we start to jump the shark. I cannot tell you how many Civil Affairs guys claim that they do Special Reconnaissance. They say this in earnest. Special Reconnaissance (SR) is a specific thing, a doctrinal Special Forces mission. It's a long-range reconnaissance patrol, a hide site, a mission support site, detailed target analysis, and a rough field routine. This is the traditional SR that everyone should think of. Green Beret traditional SR. Tradition.

Now I'm not arguing that Civil Affairs isn't SOF. They are. And I'm not arguing that Civil Affairs doesn't conduct any reconnaissance. They do. But Civil Affairs does not conduct SR. What they might do is better described as civil reconnaissance. Why not just be proud of what you do? Do it well. Civil Affairs can be hard work. It is a valuable asset in our arsenal. But it always becomes about representing yourself as something that you are not instead of just being great at what you really do. That's just a singular example from the Army. Go ahead and look into the identity crisis that the US Air Force Air Commandos had with their Special Operations Weather Teams becoming Special Reconnaissance. Gentle reader, USAF weather guys, who always referred to themselves as Special Operations Weathermen (because that's what they were officially called) were widely derided for their insistence that they be recognized as special operations.

To be clear, there was nothing special about the way that Special Operations Weatherman forecasted and reported the weather. Understanding the weather and its impacts on the mission is a critical component of effective mission planning, to be sure. But the almost boastful way that they constantly reminded you about the *special operations* part was awkward at best. Thankfully, when the Air Force made the name change, they also significantly updated the training pipeline to include Military Freefall and Combat Diver Qualification courses, among other substantial changes that does provide some real credibility to the *special* part.

One of the grossest cases of this inflation of self and mission for the sake of inflation was a Psychological Operations guy who was attached to a Special Mission Unit (SMU) in Syria. We know who he was, where he was, and what he was doing because he openly posted online about it. There is no remaining evidence of his escapades, but at one time he was a prolific Instagram poster. At one point he had taken possession of a captured enemy sniper rifle. He promptly assigned himself the role of *SMU Designated Marksman*. It was in his posted bio. He showed us plenty of pictures of his excellent overwatch positions and his shemagh scarf draped across his kit and his conveniently obscured face. He was certain to include pictures of his 'SMU teammates.' He made it very clear that he was mission critical, and his skills were the difference between success and failure on the cutting edge of the battlefield. Let me be clear; there does not exist, ever, a set of circumstances where an untrained PsyOp soldier is given the task of sniper overwatch for a Special Mission Unit in Syria. Or anyplace else. His exploits garnered some attention in certain circles and his profile vanished. I hope that he got all the attention that he wanted. He deserved it.

But this distorted understanding of roles and responsibilities is, I believe, a logical and inevitable conclusion of the distortion of the language surrounding the SF vs SOF vs lowercase special forces discussion. You should begin your journey into the world of SF…capital S, capital F…understanding and reinforcing the correct language. Your language influences your thinking. As your journey continues and your language evolves you will have already developed the proper operational phonetics and your understanding of the language will enhance your learning rather than inhibit it. It is always best to be fluent when you are talking shit.

No Shit, There I Was...Special Reconnaissance

Have you ever shat in a bag? During the Qualification Course I was witness to one of the most absurd and funniest shitting in a bag stories ever to be told.

I had a good friend, let's call him Brad, who was an experienced reconnaissance man. He had served in the 82nd Airborne Division's Long Range Reconnaissance and Surveillance Company, known as LRRS-C. During his tour there he was able to attend Military Freefall School, so he had already distinguished himself amongst an impressive group of Officers. Brad was an excellent Soldier and went on to serve with some distinction, but he also had a mischievous side and was quick to take advantage of the credibility he had earned for his service with LRRS-C. He was rightfully seen as a subject matter expert in all things reconnaissance.

During the Q course, we were partaking in a detailed Special Reconnaissance mission, including a full 96-hour isolation and planning routine and the subsequent mission execution. Brad became the source for much of our collective understanding of this complex mission set and we learned to trust his judgment as solid, nearly unassailable. Blind trust almost. You can see where this might lead.

After the requisite planning, back briefs, and rehearsals at Fort Bragg, we conducted our airborne insertion and moved to our mission support site deep in the woods of Fort Pickett, VA. Our team of merry Captains went about the laborious task of establishing area security and creating a hidden hide site near our designated target. This little hidey-hole ended up being a small copse of evergreens a few hundred meters away. We diligently established a rotation schedule so that three guys moved from the support site to the hide site during every period of darkness. Under the watchful eye of senior Special Forces NCO Cadre, the tour of duty in the hide site was one of strict noise, light, and movement discipline. A hard routine to be certain. We were serious men learning our serious craft. It was also supremely boring.

Brad and I ended up on a hide site rotation with Rob. Rob was also an excellent officer, but a little prone to hysterics. Brad, sensing an opportunity, took advantage of his exalted position and decided to have a little fun. I happened to be manning the optics, so I was mostly an innocent bystander. Mostly. Brad turns to Rob and whispers:

"Hey Rob, I need you to hold my bag."

"What do you mean, what bag?"

Brad responds, "I have to take a dump, and I need you to hold my bag." He went on to explain that the technique was to take an empty heavy-duty MRE bag, carefully fold down the sides, relieve yourself in the bag, and then carefully roll up the bag and seal it with tape. You then pack out your waste leaving no evidence that you were ever near the target. Tread lightly, SR style.

Rob immediately objects. "No way, man! Hold your own bag!"

Brad asserts, "That's not how it works. Your buddy has to hold your bag. Trust me, I was in LRRS-C."

Pointing at me Rob insists, "Why can't he hold it?!"

"Because he's on the optics. Stop being such a wuss. Just help me out. I know what I'm doing." Brad emphasizes.

Brad preps the bag and shoves it at Rob, who reluctantly assumes a prone position. Mind you, this is all taking place in a copse of trees just slightly larger than the bed of a pickup truck just a few hundred meters away from our very serious target.

Brad slowly wriggles his pants down and grabs on to one of the trees in a deranged stripper pole stance, leans back, and drops his ass down just inches from Rob's waiting bag below. I should note that we have been in the field for some time now and we weren't exactly fresh in the undercarriage.

Rob immediately protests. In an exaggerated stage whisper, he gags out, "C'mon man! I can't do this!" Brad breathlessly whispers back, "Hold the bag still or I'm gonna shit where we are laying!"

Rob gags back, "Hurry up…gag…I can't do this…gag!"

Brad, sensing Rob's dismay, turns up the heat and starts swinging a little bit. Bare ass, field funk, and a panicked bag boy.

Rob is really starting to lose his cool. Gagging. Pleading. It's beautiful. I'm starting to lose my very serious composure. Brad swings, Rob gags, rinse and repeat.

Brad goes in for the kill, drops his balls down, and actually drags them across Rob's knuckles.

Rob loses it and squeals in disgust. Literally squeals.

And then we see the Cadre standing right next to us. In the chaos of the evergreen stripper, we lost our situational awareness and the Cadre just happened to come and check up on us. We are sunk.

I can see my Special Forces career vanish right before my eyes. The Cadre stare in that special angry way that only seasoned NCOs can. Brad slowly stands and pulls his pants up. Rob, thankful for the respite, breathes a sigh of relief. The Cadre stare. We shrink.

Then the Cadre bust out laughing. Full on belly laughing and guffaws. They have been witness to one of the greatest Q course trolls of all time. They can appreciate great art.

We were gently admonished, but we can see the admiration on the Cadre's faces. Our relief is palpable. We settle back into our very serious duties, and I turn to Brad and ask him, "Do you still have to take a shit?"

He turns to me and nonchalantly replies, "I never had to."

And that's why you never trust a LRRS man.

Life Lesson: Don't take life too seriously. Even in the most monotonous or boring situations find a way to laugh, find a way to keep yourself occupied. Those around you will thank you. Find the light when darkness is inevitable.

Big Boy Rules

One of the defining organizational culture markers in Special Forces is the idea of *Big Boy Rules*. The idea is that you will be treated like an adult and not subjected to the traditional dogmatic application of military chicanery. You will hear this cited as one of the primary reasons people want to come to Special Forces units. Nobody will yell at you for uniformity or putting your hands in your pockets, you will not have to endure the drudgery of unit PT, or the slog of formation runs. You might get to grow a beard and you will generally be left alone to do your job the best way that you see fit.

As you might imagine, this culture is vastly different from the conventional military mindset. It is not uncommon for unit leaders to strip Soldiers of their autonomy. If you show up to a winter PT formation and someone forgot their watch cap, then everyone is forced to remove theirs, in the name of uniformity. Your cranky old First Sergeant will not allow you to grow a mustache even though it is well within the regulations and has absolutely no impact on mission readiness, just because he said so. You will spend your Monday, your entire Monday, in the Motor Pool conducting Preventive Maintenance Checks and Services on the one vehicle that you are assigned even though it has been deadlined for weeks on end and there is absolutely nothing that you can do at your level to fix it, simply because it is Motor Pool Monday. I have even heard of entire formations required to set up the pouches on their body armor identically, even though some guys are left-handed or have different weapons assigned to them. Everyone must be the same regardless of how it impacts combat readiness and effectiveness. Good order and discipline, they call it.

Perhaps the most egregious example of this coddling mentality is the *recall formation.* The lower enlisted folks are required to stay at work until the unit can hold the end-of-day recall formation. You will sit around, waiting on the word, until a predetermined time to stand in a formation because the training schedule that was created weeks prior says so. It does not matter that you have completed your assigned tasks for the day and the unit area is in generally good repair and your training records are up to snuff. You are not allowed to go home because the leadership might have something quasi-important to put out at an arbitrary closeout formation. This 'just in case' mentality is always at the expense of the lowest ranking members who have no autonomy and no authority. Shit rolls downhill and these poor people are at the bottom of the hill. This is the same broken conditional thinking that begets so-called 'zero-tolerance' policies that so often go astray. We have likely all heard the tale of the young schoolboy who chewed his Pop-Tart into the shape of a pistol and earned himself a suspension because the school has a zero-tolerance gun policy. The conventional Army is full of these rigid ways of thinking.

I should note, however, that some of these seemingly arbitrary rules are born of necessity. Soldiers are famous for pushing boundaries and testing the limits. If you don't draw the line somewhere, someone is bound to steal the markers. There is a reason why the desiccant packet in the MRE says *Do Not Eat* and the M18A1 Claymore Mine is so obviously labeled *Front Toward Enemy*. If you don't enforce some semblance of uniform standards, it will not be long before guys start coming to formation in all manner of clown shoes, metaphorically speaking. So Big Boy Rules is the desire, but Little Kid Rules is the norm in the conventional Army.

In Special Forces, the norm is Big Boy Rules. The problem is that most people ignore the second part of Big Boy Rules, which is Big Boy Consequences. You are not likely to be corrected for minor uniform infractions like your pants unbloused from your boots or non-standard PT uniforms. Hands in pockets is not just tolerated, it is anticipated. However, the expectation is that you will not walk around in cut-off service uniform pants or go to the PX in a mixed uniform wearing your ODA sweatshirt giving some conventional Sergeant Major a target of opportunity. You will not be subjected to mass PT formations, but you are not allowed to pencil whip PT Cards or walk around with a beer gut. You can run whatever accessories you would like on your guns, so long as it serves a functional tactical purpose. The icing on the cake is that when the work is done, even if it is done at noon, you get to go home. Sweet, sweet freedom. You earn a lot of sovereignty, but the expectations of your performance and your behavior are much higher.

The consequence part is two-fold. One of the consequences of the Special Forces environment is that you have to exercise more discipline so you can enjoy more independence, as discussed. The other, less often discussed element of Big Boy Consequences is the actual consequence part. The penalties part. The punishment part. In this regard Special Forces has not done a very good job of late. And 'of late' is doing a lot of work in that statement. Special Operations in general, and Special Forces in particular, has had a spectacular run of legal, ethical, and moral challenges that have blackened the collective eye and have made substantial withdrawals on that credibility bank account we discussed earlier. It is a SOF Imperative for crying out loud! *Ensure legitimacy and credibility of Special Operations.* Special Forces has done so much damage to our credibility that there are places where we are not welcome. Ambassadors and Commanders are predictably reluctant to allow SF into their countries and areas of operation. We did this to ourselves. We can undo it.

I'm not certain that I have all the answers to this conundrum. I do have some of the answers, I believe. There is a recruiting and assessment component, an institutional training and education component, and a leadership component. There should also be a punishment component. There must be. It is inevitable. When a Green Beret blatantly and wantonly violates his obligation to the Regiment, the Regiment should respond in kind. In grave legal cases we should allow the justice system to work. When a guilty verdict is found we should follow that with a total banishment. We revoke your tab. We take your Green Beret. We kick you out of the Army. When a moral obligation is deliberately abandoned or you forsake our ethical values, we will take your tab and your beret. And we should do so publicly. The Special Forces community is a great deal reliant on reputation, both inward and out. You trade on your name long before you walk into a room. Your bona fides, what people think about you, is a large part of your status in the group dynamic. As such, when you fail to uphold the standards of conduct then we should let it be known far and wide that you are no longer part of the in-group. You are out.

The misadventures of the Brotherhood are fairly well-known. I should note that the bad apples are genuinely very rare. The overwhelming majority of Green Berets are decent and honorable and even paragons of virtue. However, when a Green Beret messes up it makes the news. What is almost entirely obscured from the public eye is what happens to that Green Beret when the dust settles. How he was punished. What the consequences of his actions were. I predict that our hand will soon be forced. Our infrequent misdeeds have become so dire, not necessarily so common, but so dire that the option for discretion will be revoked. I think this is a good thing. So the takeaway for the Special Forces Candidate is that you should be prepared to make your way in this new environment. That is another one of the SOF Imperatives;

understand the operational environment. We need to accept that with Big Boy Rules come Big Boy Consequences.

We are already seeing this culture of accountability manifest in SFAS. Minor infractions are not tolerated. All the rules matter. If you are briefed that you must not be farther than an arm's length away from your ruck or your weapon and you lose your situational awareness and violate that rule, then be prepared to be dropped for not meeting the standards. If you are told that your ruck must weigh 50 pounds and your ruck only weighs 49.9 pounds, then your ruck doesn't weigh 50 pounds and you get what you earned. If you are told not to walk on the roads during land navigation, then stay off the roads. The message that was sent to my SFAS class in 1997 was clear. You will draw a picture of the patrol cap because it represents selfless service to the Regiment. The message at SFAS now is equally clear. All the rules matter, and the standard is the standard. The rules matter so much and the consequences are so extreme because when the shit hits the fan it is usually more than just your reputation on the line. We want you, but we only want you on the terms that we set. Those terms are ones that will set a culture of accountability and elite performance. Now and for generations to come. This is the type of organization that you want to be a part of.

The 3 Rules

1- Always look cool.

2- Never get lost.

3- If you get lost, look cool.

The origin of the 3 Rules is an oft debated topic. They are sometimes attributed to Explosive Ordinance Disposal guys from the WW2 era, but EOD guys are not all that special. Crazy, yes. But not special. The SEALs sometimes take credit for establishing the 3 Rules during the Vietnam era. The SEALs have a dubious history of accountability so let me check that overreach now. Some sources cite the 3 Rules as belonging to all Special Operations Forces. But I have seen some of the most misguided map and compass debacles at the hands of special forces, lowercase s, lowercase f. So, I am going to stake the official claim right here, right now. The 3 Rules belongs to Special Forces, the Green Berets.

If you want to challenge this claim, then write your own book. But you will not look very cool doing so and that will invalidate your claim from the start. Checkmate. Now that we have established who rightfully keeps the 3 Rules, we should spend some time understanding them. They are not what they may seem. They are not about style points, or navigation skills, or at least not entirely. The 3 Rules are a deceptively complex mental model that can be applied to most of life. It can certainly be applied to SFAS. In fact, it may be the most critical framework that you can apply to successfully get selected. Pay attention.

Always Look Cool. There is something inherently attractive about a well put together uniform. The regalia of the military man is a time-honored and respected look. The venerable Triple Canopy: the stacked Special Forces, Ranger, and Airborne tabs on the left shoulder is duly respected. It just looks cool. It is cool. I acknowledge that even though I spent some time explaining how important it was to understand the symbolic distinction of the Green Beret, there is no denying that it does look cool. Looking cool is so imbedded in our awareness that we have a loosely established, but widely understood and accepted, set of rules about what is and is not suitable dress for our civilian attire.

I often joke that it is relatively easy to pick out Green Berets from the crowd. They almost always have a Suunto or Garmin watch, likely set on dark mode. They are probably wearing Merrill or Altima shoes, although sometimes a pair of Vans will slip into the line-up. They prefer high-end tactical casual khaki pants that don't look too *operator* but will allow for some on-the-fly fire and maneuver. Something from Lululemon, Arc'teryx, or prAna is always cool. Carhart and 5.11 are acceptable, but a little too on-the-nose for some. The term 'Patagucci' was coined at Fort Bragg. They likely have a GoRuck or Mystery Ranch bag slung over one shoulder and unless you are at the airport, they always have a discreet little combat folder tucked into a slightly frayed front pocket. Throw on some ubiquitous pair of dark wraparound sunglasses and you have the acceptable facsimile for your average Green Beret role-playing the grey man[2]. Of course, you have to look fit as well. I know there is some Green Beret out there reading this right now and he is mentally scrolling through his wardrobe and knowingly nodding at the accuracy of this description. So, looking cool, actually looking cool, is a real thing. In uniform and out. But we are searching for something deeper.

Perhaps the most important part of looking cool is being switched on, knowing what you are doing, and

---

[2] Don't confuse this with the Grey Man concept, wherein the Grey Man does not draw attention to himself. He can disappear into a crowd and move unnoticed amongst the local population. This is a worthy concept, but it rarely applies to SF guys, particularly when moving as a group. Four 6-foot yoked gringos walking around in some city in Latin America or Asia look like four 6-foot yoked gringos…nobody is confusing them for locals.

displaying confidence. When the shit hits the fan, the panicked masses are not looking for the meek and meager. They don't need some weak functionary who waffles and deflects. When a Green Beret stands in front of a senior commander or an ambassador and briefs his scheme of maneuver, he needs to be switched on. He must speak with complete confidence and clarity. But he is not just a salesman, he must be competent. He must know the plan and it has to be a good plan. Green Berets have a long tradition of superbly refined planning skills. The traditional 96-hour Special Forces Detachment Mission Planning process (USAJFKSWCS, 2020) requires a level of mastery that allows relatively junior Special Forces Non-Commissioned Officers to rival the planning prowess of conventional Army Officers. If you have ever witnessed an ill-prepared staff officer present a poorly designed and insufficiently understood plan to a critical audience and get eviscerated for his incompetence, then you understand the importance of looking cool. Getting your ass chewed because you were not switched on enough to *develop multiple options, ensure long-term sustainment, and provide sufficient intelligence* (please see: SOF Imperatives) does not look cool.

To understand this paradigm in the context of SFAS you might consider the following: Selection is a 3-week long job interview, under the most extreme conditions that you could imagine. The interview questions are sometimes demanding physical tasks. They are designed to get you flustered. SFAS is intended to stress you, to make you lose your cool. When you lose your cool, we take note. We want you to perform, under pressure, for three weeks straight. At SFAS your briefing duties will be light, but your performance requirements are incredibly heavy. Your measures of competence are rucking and running, land navigation, and strong interpersonal skills. The scope of your required aptitude is grand by any measure. Competence builds confidence. So, always look cool.

Never Get Lost. Wise words on their face. Especially wise at Selection. Land Navigation failures account for nearly half of selection drops. Accordingly, I have provided an entire section on land navigation in this book. It is that important. On the micro scale, this advice is obvious. On the macro scale, it can take many meanings. I would suggest that the expectant Candidate emphasize two specific meanings, the obvious orienteering meaning and the deeper philosophical meaning. Your literal compass and your moral compass. Know who you are and to thine own self be true. I don't necessarily mean in the biblical sense, but your morals are certainly shaped by your faith. When you finally get to Group and you get the opportunity to deploy, you will be exposed to any number of ethical challenges. We need you to be prepared for this eventuality, but first let's get you through Selection.

    One of the key markers for success at SFAS is intrinsic motivation (Farina, et al., Physical performance, demographic, psychological, and physiological predictors of success in the U.S. Army Special Forces Assessment and Selection course, 2019). Not necessarily while you are at Camp Mackall, although that is certainly a factor. Rather, your road to Selection is often a lonely one. The reasons for this should be apparent. Adequately preparing for the physical rigors of Selection will likely take you months, and you will be expected to do most of this on your own. Your unit probably doesn't care if you are going to Selection so you will not be excused from unit PT (remember our discussion of autonomy earlier?) and as a result you will be training on your own time when your peers are off enjoying themselves. In fact, many Candidates that I interviewed reported that their units actively, perhaps not formally, but actively dissuaded them from going to SFAS. Some units see your volunteering as a slight to them, organizationally *and* individually. Comments like, "Do you think you're better than us?" or "You'll never make it," are common. This is a sad reality of human nature.

So, your time spent preparing for Selection will likely be a bit isolated and filled with distraction. The long rucks, the innumerable gym sessions, the land navigation training, and the unavoidable rehabilitation when you twist your ankle, or blow out a hammy, or jam your knee will be mostly with your own thoughts. If you haven't taken the time to steel yourself for this clash with your own demons, then you will probably lose this battle. The sort of motivation that is born of hatred for your current station in life is unlikely to yield the long-term and healthy results that you will need. In the same way that only following the rules because you fear getting caught rather than following the rules because it is inherently the right thing to do is unsustainable, so too is motivation wrongly reasoned. Know why you are going to selection. Patrol cap or beret. Know what sort of things you will be doing when you earn your Green Beret. Uppercase or lowercase. Know why you are doing this thing, committing to this lifestyle, and making the sacrifices inherent to doing so successfully. If you can set this compass for SFAS, then you will be in a much better position to read the distorted ethical landscape you will later encounter.

If you get lost, look cool. If you remember one thing from this discussion of the 3 Rules, let it be this. You will get lost. It is entirely unavoidable. I mean this both in the physical sense and the metaphysical sense as well. During your land navigation you will lose track of where you are. Maybe not catastrophically lost but lost, nonetheless. The same is true in the abstract. When the weight is bearing on your back, your legs are cramped and bent, or your feet are so badly blistered that it takes ten minutes just to put your socks and boots on, you will be lost. Lost in your own thoughts of self-doubt and imposter syndrome. You will tell yourself that you are not cut out for this endeavor. You will take comfort in failure. You will ask yourself, "Wouldn't it be nice to get some hot chow and a shower?"

Expect this lesser version of yourself. In the literal sense you must learn the skills to reorient yourself to your surroundings. Take out your map, navigate back to your last rally point, and restart that leg of your route. Have the discipline and skills to get *un-lost*. You must build this discipline in the metaphysical sense as well. Have a deep faith in a higher purpose for your current misery. Know, in a profound sense, what your effort is worth. If your only reason for going to selection is because you hate your unit leadership, you will just return to them a bitter failure. It is fine to not like them, but do not use that as your source of strength. Know thine own self and make certain that you are the type of person worth the sacrifice.

When the inevitable happens and you find yourself suddenly searching for the right path, be prepared for it. If you prepared correctly then you will have the skills to find your way, the strength to complete the task, and the mental fortitude to silence your inner critic. At Selection you will have these attributes because you were disciplined enough to do the hard work to acquire them. Later in life, when you are deployed and working with an unsavory partner or engaged in a nebulous ethical environment, this discipline will serve you well. You should build the sort of strong moral compass that Green Beret Sergeant First Class Charles Martland has. In September 2011, Martland found himself in an ethical dilemma. The Afghan partner force commander that his ODA was assigned to work with was sexually assaulting little boys. The barbaric Pashtu practice of bacha bazi is inconsistent with any of our American values and should be met with nothing short of absolute disgust and derision. This particular Afghan commander was in charge of a unit that was vital to mission success and despite repeated attempts to persuade him to see the error of his ways the behavior continued. Martland had enough, rightfully so. During a heated exchange Martland kicked his ass. He beat that pedophile like the dog he was. His actions earned him a convoluted and unreasoned disciplinary action from the Army that almost forced him from service. He was eventually spared, but not before his career was inextricably derailed. An immense sacrifice to be certain. But he will suffer no moral injury. He can hold his head high. His status among the Brotherhood is assured. He will never buy his own beer if he is with me and the list of contributors to his bar tab is as long and distinguished as is his Special Forces tab. Build that fortitude now and you will be able to call upon it in the future. At SFAS and beyond.

Unconventional Warfare

To truly understand Special Forces, you must understand Unconventional Warfare (UW). I don't mean understand the doctrinal definition and tactical application of UW. If you want that sort of stuff then please go read Army Techniques Publication 3-18.1, *Special Forces Unconventional Warfare*. In fact, you should read ATP 3-18.1, but not yet. I mean that you must understand the way this unique mission influences every aspect of Green Beret identity. UW is not just warfare conducted in an unconventional manner. Performing a raid with some novel infiltration technique or applying an asymmetrical force multiplier to your civil engagement plan is not UW. Green Berets are the only Department of Defense entity that is specifically trained, manned, and equipped to conduct UW. The ODA is designed to be capable of conducting decentralized split-team operations without any sacrifice to capability. Every team has two medics, two engineers, and two command and control nodes. When pressed, an ODA can advise and assist an entire battalion of partner forces. Twelve men create 500 fighters.

A Special Forces team can infiltrate denied territory, link up with partisan underground forces, establish and maintain auxiliary support networks to develop intelligence, procure logistical support, and conduct a vast array of military operations. We can turn farmers into a standing army. One team, with twelve Green Berets, can give you the sort of strategic flexibility and tactical lethality that gives our adversaries pause. That's called Strategic Deterrence. You can teach a squad of cooks how to conduct half-way decent quasi-CQB room clearing in short order[3]. Those sexy videos that you see of operators kicking in doors and shooting people in the face make for decent propaganda, but they mask the real purpose of the Green Beret. Unconventional Warfare.

    This sort of eccentric combat environment requires a different breed to manage the chaos. You have to be tactically and technically sound, to be certain. Understanding small unit tactics, combat marksmanship, and first aid is a given. But what will you do if the local farmer that you depend on for secretly supplying your food and shelter has a sick animal? You better be certain that your 18D remembers his livestock medicine. What happens if your extended stay in the Republic of Pineland exhausts your supply of batteries for your radios? You better ensure that your 18E is up to snuff on his signal wave propagation and that he can construct the proper antenna to put into working order the ancient radio set that he found in the basement of a local hobbyist. What will you do when your

---

[3] Likewise, just because your squad of Engineers or MPs did some urban combat, or you spent a day on the range with some hapless ODA doesn't mean that they are ready for SF-like missions. Spend a few minutes researching MOUT, AMOUT, and CQB and understand the differences between controlled pairs and a double-tap and you'll start scratching the surface of this incredibly nuanced mission set.

weapons start to show wear from the extended firefights and protracted skirmishes that you are inflicting on your enemies? You will require the skill of a talented armorer and that aptitude should reside in your 18B. We haven't even touched on the interpersonal skills that it takes to manage the competing factions of your guerilla force, or the intuition required to walk into an Area Command meeting and understand the power dynamics and who the real decision makers are.

Unconventional Warfare requires a Green Beret to be a renaissance man. A jack of all trades with the skill of a craftsman, the eye of an artist, and the mind of an academic[4]. A tactical athlete that can outrun, outfight, and outthink anyone who opposes him. As the OSS, the ancestral precursor to the Special Forces, would say – we want PhDs than can win in a bar fight (Lujan, 2013). You will need an advanced degree in…well, everything. Even when you work outside of a declared theater of active armed combat you will need your wits about you. You will find yourself as the senior US Government representative at

---

[4] The full quote from William Shakespeare is, "A jack of all trades is a master of none, but often times better than a master of one." This aptly describes what Green Berets are and why we are often the force of choice for unknown and unknowable problems.

many functions, whether you like it or not. You are not just a trigger puller; you are a policy implementer. You are probably not seen only as the guy that can calculate the correct release point so that your freefall parachute operation lands on the correct drop zone, which you absolutely must do precisely. You will also likely be seen by your host nation counterparts as the duly appointed delegate of either the shining beacon of hope upon the hill or the Great Satan, depending on where in the world your drop zone ends up. Watch out, here come those SOF Imperatives again, *recognize political implications*. You are an expeditionary political scientist, so you better be able to wrestle with International Relations theory *and* the proper application of a choke hold. The modern Green Beret is a renaissance man, because modern Unconventional Warfare demands it. So be certain that you know what you are signing up for and be prepared for catastrophic success, you might just get selected.

No Shit, There I Was...Chemlight Voodoo

In the Fall of 1994, the US executed Operation Uphold Democracy in Haiti. This multifaceted Humanitarian Assistance, Foreign Internal Defense, and quasi-Unconventional Warfare effort included Green Berets to help provide security and stability to the beleaguered island nation. Haiti was, and to this day remains, a dangerous and often backwards social structure that defies logic. As eccentric as the operating environment is in the political and diplomatic spheres, the spiritual environment is downright crazy. Voodoo magic and witch doctors are often the informal social leaders and exercise immense control over the populace. It is not uncommon for rural areas to ignore any central governance efforts and seek all guidance from this informal network of black magic purveyors.

As you can imagine this creates some problems for an occupying force who is seeking to legitimize the standing government and provide security and stability. This is a difficult task under perfect conditions, but a shadow government of snake-gods and zombie worship makes this as unusual as you could ask for. This is exactly the sort of situation that calls for an unconventional warrior. Every zombie uprising requires a Green Beret. So, the US Commanders tasked the Special Forces units assigned to them to figure out exactly what the problem was, and more importantly to get the situation under control.

Those Green Berets set out into the countryside to evaluate the situation. They visited the villages and talked to the villagers. They met with any local government officials that were still in position and they mapped the local networks of influence and power. It quickly became apparent that these witch doctors were a malign influence and they defied logic. A cholera outbreak was besetting the country and US officials were eager to treat the populace and prevent a public health crisis on top of the security crisis. Treating cholera is straightforward with a simple course of antibiotics and rehydration therapy. But according to the resident witch doctors that sort of stuff was bad magic and a product of the failing regime.

The Green Berets on the ground worked tirelessly to convince the locals to accept the treatments at face value. No strings attached and no favors owed. Please take this medicine and be healthy, save your kids, and separate this critical care from your feelings of the legacy administration. This is not bad magic; it is not any magic. It is just medicine. Lifesaving medicine. Take it and be well.

The witch doctors rightfully viewed this as an encroachment on their power. If these white devils could deliver health, then they could also deliver legitimacy. This represented a very real threat to their continued relevance, so they fought it with all their influence and authority. Frustrated with their progress the Green Berets decided to fight power with power. Black magic with black magic.

One night, an enterprising Special Forces Team Sergeant strapped some plastic chemlights to his arms, and carefully concealed them under his uniform sleeves. With his ODA in tow, he visited a prominent remote village to speak with the witch doctor and instigate a conflict. During this heated argument, the Green Beret baited the witch doctor into a challenge of magical powers. The witch doctor chanted his chants and danced his dances, and the gathered village was duly impressed. The Green Beret stood and began his enchanted show. He recited old magic and summoned great powers. Members of his team were stricken zombie-like with his incantations. The villagers took notice. The show continued at pace with more wild dancing and screams of enchanted charms. Finally, at the peak of his performance the Green Beret slashed his arms with a magical knife and his previously concealed glowing magic blood was revealed. The zombie ODA was released of their trance and magically restored to life. Powerful magic indeed.

The new Green Beret master of the dark arts told the impressed villagers that he would share his great powers with them. If they wanted to avoid the bad juju and the zombie curse, they should accept this magical cholera medicine willingly. The choice was theirs. The team would return in the morning with more magical antibiotics. The following morning the Special Forces team arrived to a long queue of eager patients. These villagers had witnessed great magic and were anxious to be well. Word quickly spread in the hinterland and in a few short days villages across the country were similarly turned. What was once an unknown and unknowable problem became an opportunity for critical and creative thinking.

And that's why every zombie encounter requires a Green Beret response.

Life Lesson: Don't think outside the box, get rid of the box. If you are beating yourself on a technicality that you created, then change the rules. If you keep thinking the same way, you'll keep getting the same results.

Chapter 3 – The Upside-Down World

What is SFAS? It is impossible. Not impossible to complete, impossible to understand. SFAS defies the traditional military training paradigm. Every military task that you train outside of Selection consists of Task, Condition, and Standard. Do this thing, under these circumstances, to this specific quantifiable criterion. Not so at Selection, where it is simply Task and Condition. You will not know the Standard. Only a select few do know. The standards are almost unbearably high, and you simply will not know when they are met. For many Candidates this is very disconcerting, off-putting even. Candidates often wonder, "How will I know what to do, how to do it, and when I am complete?" The customary reply from Cadre is, "Do what you think is right, Candidate," or "Do your best." What does the best look like? Let's see.

SFAS has a long but fairly straightforward history. Selection was created in 1988 by Gen. James Guest of the John F. Kennedy Special Warfare Center and School (Feeley, 1998)[5]. In 1987 Special Forces became a

---

[5] There is still some debate about the actual founding of SFAS. The work to build a 'selection' started in 1986, went through several pilot and test programs, and was mostly finished by 1988. It didn't get the official name of Special Forces Assessment and

recognized branch in the Army and SWCS was eager to "provide highly suitable soldiers for Special Forces" in a greater effort to professionalize and institutionalize the force (Velky, 1990). Guest set about developing a process to do so and after several years of study and testing, SWCS created SFAS. It has remained virtually unchanged in its construct since its inception. There was a brief period in the mid-2000s when SFAS was shortened to just two weeks, but even that failed experiment didn't alter the sequence too much. It has since returned to its original structure and following a brief reconciliation process circa 2010, it has been continuously screening and selecting Candidates in a detailed deliberate manner.

Hosted at Camp Mackall, North Carolina SFAS represents one of the most physically demanding processes in the Department of Defense. Camp Mackall is a small auxiliary installation about 40 miles west of Fort Bragg that began its existence as Camp Hoffman in 1943 as part of the burgeoning US Army Airborne initiative (Hagerman, 1997). After WWII it grew to become an austere outpost that was once the exclusive domain of the Green Beret. In recent years it has become the home of several units that

---

Selection until 1989, but it was the same scope and sequence as it was in 1988, so we'll settle on 1988 as the 'creation' date.

take advantage of the attached airfield and the mostly unoccupied surrounding airspace. The Camp Mackall of today would shock the SFAS attendee from just 20 years ago. The explosion of growth and accommodations is remarkable. When I attended SFAS in 1997 we were often derided by the older Cadre because we had access to what was called the "Million Dollar Shitter." It was a crude, but recently constructed quasi open-air latrine facility. It was anything but luxurious, but compared to slit trenches and the infamous tar paper shacks of old, it was considered downright decadent. Those Cadre would be shocked to see what Camp Mackall has grown into today. The buildings are deluxe by comparison, but the essence of Camp Mackall remains unchanged. Sugar soft sand that swallows Candidates, tall pines that provide scant relief from the Carolina sun, and unwavering standards. In fact, you would be hard-pressed to find any change in the standards in today's SFAS from its inception in 1988.

SFAS Described

Broadly speaking SFAS is three weeks long, give or take a few days to accommodate travel, acclimatization, or other administrative restrictions. Selection is conveniently divided into three distinct weeks; Gate Week, Land Navigation Week, and Team Week. Let's break it down week by week. I will not dishonor my NDA and I'll simply synthesize the publicly available information. In the following chapters we will evaluate this analysis in the context of what you can do to prepare for SFAS.

The performance environment is shrouded in mystery. The training calendar is not published, and the specific events rotate to prevent Candidates from predicting precisely what will happen next. SFAS does follow a common construct from class to class and SWCS has published multiple articles, several public videos, and even endorsed fairly revealing, albeit awkward, documentaries that show the sequence in good detail. From these sources we can cobble together a shockingly accurate picture of just what to expect. Interestingly, most Candidates assert that they would prefer not to have the training schedule. There is a certain dread in knowing precisely which sledgehammer is about to slam into you. What this uncertainty creates is a requirement to be well-rounded. Smart, responsible, and fit. Faster than the strongest lifter and stronger than the fastest runner. Adaptable, resilient, and capable. Or as the OSS would put it, PhDs that can win a bar fight. SFAS is a condensed demonstration of the Green Beret mission. These missions, like SFAS, are often unknown and sometimes unknowable. Here is what we do know.

Gate Week

The first week of SFAS is Gate Week and is punctuated by three key categories of events. The first are in the cognitive domain. They include assessments like the Defense Language Aptitude Battery (DLAB), the Test of Adult Basic Education (TABE), the Wonderlic Personnel Test, and the General Ability Measure for Adults (GAMA)[6]. These are general assessments of cognitive ability and what they mean is that you have to be smart to be a Green Beret. There is a minimum Armed Services Vocational Aptitude Battery – General Technical score (ASVAB-GT) required for assessment into Special Forces and while this score seems to shift back and forth a few points every few years the general conceit is that intelligence is important (Beal, 2010). There is also a psychological evaluation. It is not particularly important which inventory is used so long as it measures the Big 5 Personality Traits: openness, conscientiousness, extraversion, agreeableness, and neuroticism. This Five-Factor-Model is universally recognized and the assessment

---

[6] The DLAB is sometimes administered at the end of SFAS, but I include it here as it often occurs as part of Gate Week.

inventory tool that SFAS uses is generally recognized as the gold-standard in the field. Candidates displaying any anomalies will undergo an interview with a SWCS Psychologist.

There is also a significant physical component to Gate Week. There is a Combat Readiness Assessment, described in some detail in the following prep section and the Nasty Nick obstacle course. The Nasty Nick is designed to test a Candidate's fear of heights and confined spaces. It also measures strength and endurance. Named in honor of Green Beret legend Colonel Nick Rowe, who was the driving force behind establishing Survival, Evasion, Resistance, and Escape (SERE) school, the Nasty Nick is a rite of passage for Green Beret Candidates. The course encompasses more than a mile of obstacles, including subterranean and high rope climbs. Any fear of heights or claustrophobia will quickly be revealed (Parrish, 2013). Timed rucks and runs round out the week.

NASTY NICK…THE WEAVER, *PHOTO CREDIT K. KASSENS, USAJFKSWCS*

The most important part of Gate Week, and perhaps all of SFAS, might be the rucking. I will spend significant time covering rucking, rucking workouts, foot care, and other useful prep topics later. Let me be absolutely crystal clear: rucking is king. Rucking performance is the number one predictor of success at SFAS (Teplitzky, 1990; Beal, 2010;

Farina, et al., 2019)[7]. The careful reader will note that some studies indicate that Land Navigation performance is the most predictive, but Land Navigation is simply rucking with an overland destination. To be clear, I am not saying that the timed rucks and runs are the most important events, I am saying the rucking performance is the most important performance measure. Every important decision that you make at SFAS you will make with a rucksack on your back. The times and distances for these ruck and run events are confidential. I will say that if you are moving at the "Army Standard" pace then you are unlikely to make the SFAS standard. The "Army Standard" is generally recognized as 15-minute miles with 35-pound rucks. You need to set the SFAS standard as your goal which as previously noted is to simply "Do your best," but better than the Army Standard. That advice sounds snarky, but it is excellent advice, nonetheless. More on this later. At the end of Gate Week, Candidate performance is subjected to a performance review. All the collected data points are run through the

---

[7] These three sources are critical for this discussion. They demonstrate with great clarity the importance of rucking performance for SFAS success, they provide performance benchmarks, and they are consistent across nearly 30 years of SFAS history. The evidence is overwhelming. Rucking is, and always has been, king.

SFAS *Gonkulator* and once the *gonkulations* are complete the judgement is rendered. Candidates whose performance meets the standards are allowed to continue. Those Candidate whose performance is substandard are removed from the course, Involuntarily Withdrawn (IVW) in the parlance of Camp Mackall.

THE NASTY NICK

PHOTO CREDIT K. KASSENS, USAJFKSWCS

No Shit, There I Was…Renegotiations

The Nasty Nick holds a special place in the heart of most Green Berets. It's not that it's significantly more difficult than other obstacle courses. It is certainly tougher than most. But the Darby Queen at Ranger School is tough, too. For me it's special because of what it represents. It's a gateway to the Regiment. And it reminds me of the absurdity of SFAS. Not absurd in a bad way, just absurd in an upside-down Camp Mackall world kind of way.

The Cadre took my class out to the course and demonstrated each obstacle to us for clarity. I don't know if it was a cruel joke on one of the boys or just our collective good fortune, but they had picked a bit of a big fella to be our demonstration expert. He had a beer gut and was obviously struggling to correctly negotiate some of the obstacles. In a cruel twist of fate, that Cadre walked with my team one day and it turned out that he was a speed demon with a ruck on. He damn near broke us a few times. But he was not built for obstacle courses. His struggles may have made us underestimate the effort required for success.

We started the course at the 30-foot rope climb. Climb up, touch the top, and climb down. I recall one of the Candidates really struggling. On the first obstacle. Not a good start to the event. Every time he failed the Cadre would address him in that signature deadpan manner; "Roster Number 123, you failed to properly negotiate the 30-foot rope climb. Do you wish to renegotiate the obstacle?" It wasn't a request. The fledgling Candidate would make an ever-decreasing attempt. Rinse and repeat. Finally, the Candidate braced himself and struggled his way to the top. When he reached his hand out to tap the peak, he slipped and plummeted to the ground. Thirty-foot drop, straight onto his back. He hit like a sack of potatoes.

We thought he was dead. We all watched in stunned silence. He slowly started to wiggle his arms and legs and began wheezing. Alive, but defeated. Exhausted from the effort he lay on his back catching his wind. The Cadre casually walked over and placed his feet on either side of the Candidate's head, leaned over and checked his clipboard, and stage whispered to the prone Candidate, "Roster Number 123, you failed to properly negotiate the

30-foot rope climb. Do you wish to renegotiate the obstacle?" Completely straight faced.

Only your performance matters. Not Effort. Performance.

Life Lesson: Skills matter. Climbing a rope with just strength is possible, but technique works every time. Learn useful skills and you'll be a useful person.

The Nasty Nick - 30-Foot Rope Climb

photo credit K. Kassens, USAJFKSWCS

Land Navigation Week

Land Navigation week is exactly what it says it is. Candidates will be moved to a field location in the Sandhills of North Carolina and establish a base of operations from which to conduct a series of land navigation practical exercises, both day and night. Candidates do get an excellent series of classes and ample practice runs, but it would be foolish to show up to SFAS and not have a very strong set of land navigation skills. This is one of the likely causes of many of the SFAS non-selection numbers. We know that the Land Navigation failure rate is around 35-45% (USAJFKSWCS, SFAS Daily Summary, 08-22, 07-22, 02-22). SWCS relies on Candidates being minimally prepared by other training sources prior to showing up at SFAS. All the 18X population attends 11-series OSUT where they ostensibly get adequate land navigation training. The in-service enlisted soldiers are expected to be prepared at their operational units. This is not the reality. Infantry OSUT is woefully inadequate, even with the expanded 22-week OSUT model. Soldiers might have only three or four opportunities to practice land navigation at Fort Benning. The institutional intent is they get introduced to land navigation and then that introductory exposure is reinforced at their units. It's a logical system, but 18Xs don't have a unit, they go straight into the pipeline. The Special Forces Prep Course (SFPC) is designed to be that first unit and provide that training, but SFPC is not a teach, coach, and mentor model. SFPC seems to be a smoke session disguised as a prep course.

This is not a conducive learning environment. I am not criticizing SWCS necessarily, I'm simply pointing out the prevailing observation of the prep program[8]. I have tremendous respect for the SWCS Cadre, and we must recognize that they deal with pressures from multiple masters that make simply running any SWCS course a monumental task. But the evidence is very clear that Land Navigation is becoming an elephant in the room. I'm not advocating to lower the Land Navigation standards one iota; we just need to get better land navigation training across the DoD enterprise. Officers do better at SFAS in comparison to the enlisted Candidates and one of the likely reasons is that every commissioning source has a large land

---

[8] I have had countless 18X Candidates independently confirm this observation. You would struggle to find any 18X Candidates who have a different viewpoint. Several Candidates even shared a comment made by a senior SFPC trainer, "Why would we let you do night land nav when you guys are getting lost during the day?"

navigation training component, as well as a focus on physical fitness.

The land navigation at SFAS isn't necessarily inherently more difficult, but there are several factors that play into its high effort nature. First is the rucksack. You will recall that I stated earlier that rucking performance is the most important performance measure at SFAS. Every important decision that you make at SFAS you will make with a rucksack on your back. This is especially true during land navigation. The navigation itself is fairly easy. But when you add to the navigation the 70ish pounds of your rucksack it changes things. It will alter your pace count and it will alter your mentality. There is something about the pressing weight that pinches your shoulders, sways your back, and buckles your knees. It will change the way that you make decisions. Without a ruck you might take a longer, more easily navigable route. With that ruck pain distracting your better judgment you will make poor decisions. And the pressure is continuous. By the time that you reach the STAR exam, the culmination land navigation test, you may have put on over 100 miles on your back, legs, and feet. You will be fatigued…and you still have miles to go.

LAND NAV WEEK, 70 POUNDS AND MILE AFTER MILE

PHOTO CREDIT K. KASSENS, USAJFKSWCS

The land navigation points are fairly standard for military land nav courses. Traditional picket stakes with the self-correcting grid coordinates attached on a label are the norm for most of the practical exercises. The chemlights used to mark the points for night navigation are the small type, but the points are exactly where they say they are.

The STAR exam uses point-sitters to man the points. These point-sitters rotate between contractors, out of cycle Q course students, and the odd SWCS cadre. At the STAR points, these sitters maintain an austere little campsite with poncho shelter and maybe a campfire. Some fires are larger than others and some shelters are constructed with stealth in mind. Be prepared to search.

The Sandhills surrounding Camp Mackall are a pleasant, gently rolling, and sparsely wooded setting. The draws and swamps are another story. Draws are low ground where the terrain slopes upward on each side and toward the head of the draw. They are thickly vegetated, swampy, and they are deceptively aggressive. It is not uncommon for Candidates to plot a point with an intervening draw along their route that might normally take 30 minutes to traverse, only to enter the draw and emerge on the other side 4 or 5 hours later. A single poorly plotted route can cost you the entire exercise. The swamps of the Sandhills are littered with lost maps, missing equipment that wasn't properly secured, and broken dreams. Stay out of the draws.

Land navigation as a skill is straightforward. Land navigation in application is a challenge. Land navigation in practice at SFAS is a particular challenge. The ruck plays a part, the terrain – particularly the draws – plays a part, and the time plays a significant role. The STAR course, named for the image of a star that would manifest if you drew your crisscrossing routes on the map, occurs half in the dark and half in daylight. Many Candidates, lacking the skill or confidence, wait or move too slowly at night and are too pressed for time during the daylight. As a result, they often panic, abandon the fundamentals of navigation they were taught, and the result is predictable.

There are also rules that Candidates must follow. Don't talk to other Candidates while on the course. Don't stop and rest while on the course. Candidates aren't permitted to be on the roads or trails and Candidates caught by the ever-watchful Cadre are called *roadrunners* or *roadkill*. Roadrunning is often not done out of malicious intent. Few Candidates want to cheat. What normally happens is that Candidates, having abandoned the fundamentals, entered a draw, or otherwise waste valuable time and are presented with a moral dilemma. Do I run the road to save valuable time, avoid troublesome terrain, and make my target time or do I risk sticking to the woodline and not passing the course? They made the decision, and they should pay the price, but no malevolent intent should necessarily be judged. Another rule is that no GPS devices are allowed. It sounds like a fairly obvious and completely reasonable restriction, but there are more than a few Candidates who report to SFAS with one in hand. All the rules matter.

Some more mundane rules will prove challenging as well. Candidates are often instructed to not be more than an arm's length away from their ruck or weapon while on the course. It's a simple rule and it is easy to follow. But every class it seems that some enterprising Candidate decides that they need a rest, they drop their ruck and lean their weapon on it, and step away to reorient themselves along their route. Maybe just a few paces, maybe more, but the rule is clear. No more than an arm's length away. When they get caught you will hear every excuse in the book. I didn't know. I wasn't *that* far away. I needed to rest. Rarely will you hear "I take responsibility for my actions." All the rules matter. All the time.

HOW FAR AWAY ARE THOSE WEAPONS?
PHOTO CREDIT K. KASSENS, USAJFKSWCS

Candidates are usually given multiple chances to pass the STAR course. If you pass on your first attempt, then Land Navigation Week is essentially complete. You can begin your rest and recovery in preparation for the next hurdle; it is a significant hurdle. Candidate's that are required to retest the STAR are usually given the opportunity to do so. But it is more miles on your now exhausted body and additional burden on your fatigued mind. Injury rates during land nav are significant.

Abrasions and blisters, especially on the feet, are the most common reported injuries and they are not to be taken lightly. Forty-Six percent of the reported injuries at SFAS are blisters and 44% of the injuries start at land navigation (Knapik, et al., 2019). You would be well-advised to carefully read and follow the foot prep and care section later in this book. Sustaining a small blister during land navigation and allowing it to fester untreated will likely manifest into a painful and debilitating wound by Team Week. The same for your hands. Wear your gloves when negotiating thick vegetation so you don't get little cuts that effect your grip. Team Week is hard enough when fully capable, negotiating it injured is almost impossible. Let's see why.

No Shit, There I Was...Jurassic Park

The Sandhills of Camp Mackall are generally forgiving. Gently rolling hills and not too much thick vegetation. Smart Candidates can usually plan good routes that keep them mostly free from too much trouble. Cadre have given names to key locations. Scuba Road, Puppy Palace, and Six Points are well known. And so is Jurassic Park. But it wasn't always known as Jurassic Park. I was there when it earned that name.

Scott was a Candidate in my class. He was a bit gangly but was a strong performer. You could pick him out from a distance because he was tall, he was as pale as they come, and he had the signature birth control glasses of old. He was hard to miss with his distinctive loping gait and characteristic look. One day, during Land Nav Week he was out navigating a particularly nasty bit of terrain. He came across a body of water with a road crossing it. Candidates aren't allowed to use the roads, but Scott had an idea.

As it was later described to us by a Cadre who was watching the road for Candidates looking for illicit shortcuts, he spotted Scott in the woodline on the far side of the water, just off the road. Scott, mindful of the restrictions, poked his head out of the trees in a perfect imitation of the cunning raptor antagonists from Jurassic Park. Raptor Scott slowly peered out, surveyed the scene, and popped back into the concealment. Clever Girl. The Cadre watched, certain that he would soon see this wayward Candidate on the road.

A few minutes pass. The waiting game begins. The Cadre spots his raptor-Candidate again. Slowly Scott pops his head out and surveys the scene. Long goofy neck swinging his birth control goggles across the landscape. Suddenly and to his total surprise, the Cadre watches as Scott bursts from the woodline wearing what looks like fishing waders. Scott, displaying the sort of adaptability that we look for in SF Candidates, had taken large black trash bags, looped the tie strings through his belt, and then taped the tops of the bags to his thighs. With an exaggerated ungainly lope, Scott is running through the water. Like a crazed dinosaur fleeing from a predator, raptor Scott is walking on water.

The shocked Cadre runs to intercept him at the shoreline. He forced his way through the thick vegetation and when he finally catches him, he interrogates the wayward dino-candidate. To his credit, Scott calmly explains that he wasn't on the road. In a cunning attempt to keep his boots and socks dry, he had simply fashioned makeshift waders out of trash bags and had followed the rules of the exercise.

Technically, he was right. All the rules matter, and Scott had followed the rules. Albeit with a little more style than most. The Cadre was so overwhelmed with his bravado that he retreated back to his truck in impressed silence. He radioed back to the basecamp and conferred with the other Cadre. No rules had been broken. Continue to train. Scott had made such an impression with his dinosaur moves that the terrain now bears the name Jurassic Park.

I think if you can make that sort of impact, then you deserve to get Selected. Scott will retire this year with his Green Beret and Jurassic Park lives on to claim more Candidates. Clever Girl.

Life Lesson: Understand the rules and find the gaps and seams. Exploit those openings to your advantage. Find opportunity where others find barriers.

A TYPICAL TEAM WEEK LOW CARRY EVENT

PHOTO CREDIT K. KASSENS, USAJFKSWCS

Team Week

Team Week is the essence of SFAS. Images of Candidates dragging apparatus across Camp Mackall are what most people conjure in their minds when Team Week is discussed. But you only get to Team Week if you can perform. You have to think back to Gate Week. Your ruck and run times have to be good. You have to master the Nasty Nick and the CRA. You have to be smart and test well. Then you have to negotiate Land Nav Week. The countless miles, the injuries, and the fatigue. Only if you can prove your mettle in the first two weeks of SFAS do you earn the privilege of the abject masochism of Team Week. Words can do no justice of the experience, but I'll do my best.

During Team Week, Candidates are divided into training teams usually between ten and twenty Candidates. Under the careful watch of the Cadre, teams are assigned tasks of carrying heavy loads across varied terrain, for time. The heavy load is sometimes a simple ammo crate, telephone pole, or sandbags. Sometimes the load is an apparatus that the Candidates build from a designated list of supplies. Heavy pipes, old jeep tires, and tubular nylon webbing are the tools of the trade. Sometimes they are well built, and Candidates are spared too much misery. Sometimes they are poorly built, and Candidates are forced to half carry, half roll, rebuild, restart, and bicker amongst themselves for maximum misery. Either way, misery is assured.

Let me describe what this looks like in reality. I want to describe, in context, an event that has become well-known. It appears in official SWCS messaging, official Army messaging, and various other publicly available sources. You can readily see images of this event all over social media sources. So, I'm confident that I'm not releasing anything that hasn't already been officially released, but I have a unique context that I think will be useful for understanding just how hard SFAS can be. This event embodies the best parts of Selection and exemplifies the incredible struggle that Green Beret hopefuls face. I give you *The Sandman*.

The Sandman is the nickname for the event that is officially called the *Downed Pilot*. In this event, the Candidates are issued a duffel bag packed with filled sandbags. This is the simulated downed pilot, injured during ejection, and requiring the team to carry him to safety. He weighs between 300 and 400 pounds. The team is also issued some stout metal poles and various nylon lashings, and they create a frame of sorts from which to suspend the injured pilot and carry him back to the safety of friendly lines. Once constructed, this frame rests on the back of the already heavy rucksacks and creates a total load somewhere in the neighborhood of 300 pounds per Candidate. Hundreds of pounds of dead weight suspended between four hapless Candidates linked together with rigid metal poles making each stumble transfer across the team. Each team is usually charged with two or three apparatus. But there is more.

THE SANDMAN, PART I – THE APPARATUS

PHOTO CREDIT K. KASSENS, USAJFKSWCS

Each team is also responsible for moving a disabled jeep. The jeeps are a staple of SFAS and while they do roll, they do not roll well. The decrepit jeeps, despite the decayed appearance, are actually seen as a bit of respite by most Candidates. They don't require any construction and they really roll. They are often a welcome sight, and they begin The Sandman as such. The jeeps are heavy, but in relative terms to the pilot sandbag apparatus burden, the jeep is an easy effort to move. In this case, the jeep soon becomes the albatross around the Candidates neck that provides no escape from desolation.

THE SANDMAN, PART II – A DISABLED JEEP

PHOTO CREDIT K. KASSENS, USAJFKSWCS

The Sandman task is to move these downed pilots and this disabled jeep over a predesignated route within a predesignated time frame. On its face, the task is straightforward. The distance seems reasonable, the weight seems manageable, and the time limit seems realistic. Carry this heavy object across varied terrain for time. But like most things at Selection the devil is in the details. Candidates begin the route and figure out a work cycle and rest rotation switching between pushing the jeep, carrying the apparatus, and just walking. Move a hundred meters, rotate positions, keep going. Then the weight of the apparatus seems to grow with every step and Candidates quickly learn that work position designated to push the jeep along is the preferred job. The Sandman is heavy and unforgiving. The team begins to slow as they move along the route and doubt starts to creep in. Candidates start to call for a rotation or an outright rest break. The rest breaks become an event within themselves as lifting the heavy apparatus off the rucks and placing it on the ground and then lifting it back up after the break is over becomes a herculean effort. And then Camp Mackall gets a vote.

THE SANDMAN - THE MISERY IS JUST SETTING IN...
PHOTO CREDIT K. KASSENS, USAJFKSWCS

You may recall my earlier description of Camp Mackall included a tangential mention of sugar-soft sand. That sand is important to SFAS. I don't know if the sand is a condition of the timeless geology of the region or if it is a result of countless Candidates sweat and toil, but that sugar sand is relentless. It swallows effort. Feet sink to the ankle with ease and the simple act of walking becomes a struggle. Add the burden of a rucksack and the effort becomes staggering. Literally staggering. And here comes the jeep. Wistfully rolling along, followed by the besieged apparatus. That jeep rolls right into a little sugar sand and promptly comes to a stop. It usually happens at the bottom of a small incline. A meager little hill that you normally wouldn't even take note of. But now, with your back compressed, your legs bent, and those jeep tires firmly stuck that tiny little rise seems like a mountain. And you must climb it. With dead weight and a broken jeep. And the clock is ticking.

Within a period of just a few minutes Candidates go from a reasonable distance, a manageable weight, and a realistic time limit to completely trapped. The jeep is stuck, and what moments prior, was the difficult job of carrying the Sandman is now the easy job. The once easy job of pushing the jeep is nearly impossible. Chaos ensues. Even well-coordinated teams start to break down. The signature creaking of a load-laden rucksack strap becomes a sharp squeal. The quiet determination of a once calm Candidate become frenzied calls for relief. It takes half the team to move the jeep just a few feet at a time. Progress is measured in inches. And the clock never stops. The Sandman never rests.

THE SANDMAN TRAFFIC JAM BEGINS
PHOTO CREDIT K. KASSENS, USAJFKSWCS

You can see the quiet realization in faces of the Candidates. You can see the despair in their eyes. I'm not being hyperbolic in my description. It is an emotional event to observe this transformation. I have watched this dozens of times, on this specific event, and it can be heartbreaking. You will witness grown men go from motivated hard-asses to openly crying in minutes. Eyes red, noses running, and spirits broken. They can look up that small hill and see their future. They don't want to look at the reality of their predicament. There is more than one team. Multiple teams clambering up the same road. A traffic jam of jeeps and apparatus. It is utter bedlam. And the Candidate is just standing there, with 300 pounds pressing him into the sands of Camp Mackall. Once you experience The Sandman, as a Candidate or an observer, you will never forget it.

This is just one event, on one day, of what promises to be the longest week of your life. Team Week used to be two events a day for five days. You knew that you only had to negotiate ten total events. The modern Team Week is only four days, but teams might be subjected to four events a day. You might have to do sixteen Sandman-like events. When I attended SFAS it was rare to see a Team Week day last longer than ten or at most twelve hours. In modern SFAS, it is not uncommon to see a team go twenty-two or twenty-three consecutive hours, return to the cantonment area, set down the apparatus, change their socks, and the new day begins. No rest, a cold MRE, and the inevitable struggle of another apparatus, another ammo can, and another sandbag are the only things waiting for you. At some point the teams will conduct peer evaluations where Candidates will rank order each other's performance and the team assignments will shuffle so that Cadre can watch Candidates perform with different personalities.

The modern assessment environment also presents a more challenging atmosphere than days of old. The old SFAS Team Week found Candidates assigned to leadership positions. This was inherently stressful, but for only one Candidate and for only one event. If you were not in the assigned leadership role you essentially just had to carry the heavy weight. The modern Selection employs a leaderless environment. No one is assigned as *the* leader, so everybody is *a leader*. There is never a moment when you *just* carry the heavy weight. This allows Cadre to make much more exquisite observations of the interpersonal dynamics. Nothing is missed and Cadre are adept observers. That is Team Week.

Candidates also have this misguided impression that they should strive to be the *Grey Man* at SFAS. That is, they can fly under the radar and not undergo any undue scrutiny. There does not exist a moment at SFAS, particularly during Team Week, where you remain unseen. All the scrutiny is decidedly due. Whether that scrutiny is obvious to the Candidates or not, the Cadre maintain an unblinking eye on everything that occurs. This is particularly true during Team Week. Cadre have the entire day to just watch. They watch Candidates plan the assigned task taking note of how Candidates communicate and interact with each other. They critique how Candidates take instructions and give instructions. They watch how Candidates process criticism and provide feedback. Special Forces is no place for thin skins or abrasive personalities. Cadre watch for how hard Candidates are working and they can see more clearly than you could imagine how hard someone is actually working. Not how much noise they make or how dramatic they are acting but really doing the work. Whatever Cadre might miss, which is precious little, becomes public knowledge during the peer evaluations. If you do escape the unblinking Cadre eye, your fellow Candidates will never miss the opportunity to highlight your poor performance. If three or four Candidates report similar experiences, Cadre can focus on the activity for observation and assessment. Cadre may have blinked, but they see you now. There are no Grey Men at SFAS.

No Shit, There I Was..."I'll take the hit"

In the never-ending search for "What are the Cadre looking for?", I thought I knew the answer. My entire SFAS class thought we knew. Until we didn't. We knew Mike. Mike was another LRRS guy, but he was in the 10th Mountain Division LRRS and he was an absolute stud. He was quiet and unassuming and super friendly and approachable. Quietly competent and fit as hell. He finished in the top 3 of every event we had and looked like he was doing it with ease. He scored a perfect STAR course and was a great teammate. Everyone wanted to be on his team because it pays to be a winner. Mike was a winner.

I recall the event in question very clearly because I happened to be standing right next to Mike when it happened. He wasn't on my team, but our teams were neck and neck that day. Mike was assigned as the team leader. He had issued his tasks and was overseeing the apparatus construction. A fairly simple 500-pound wheeled apparatus and some water cans. It was a short movement, but the time limit was very fast. Our teams happened to start our movement at the same time and one of his teammates noticed that my team was only resting the water cans on our rucks. Mike's team had strapped the cans on theirs, in violation of the event instructions.

One of Mike's teammates asked out loud if they should remove the straps and carry the cans in the proper configuration. Mike, not recognizing the peril and noticing my team pulling away, said "No, don't worry about it. I'll take the hit." We completed Team Week and the final Long-Range Movement. Mike, of course, finished first with hours to spare in front of the exhausted pack. We were feeling the weight of the last three weeks, but we knew that we left it all out on the trail. The work was done and now it was time to see how we had been assessed. Select or non-select.

The class was called to the large classroom, and we shuffled inside nursing our bruised and battered bodies. A Cadre walked to the front of the room and with the signature detachment we had come to expect announced, "If I call your roster number, grab your gear and move to the back of the classroom and make a formation outside." We didn't know if he was calling the selects or the non-selects. The tension was nearly unbearable. We had started SFAS with almost 350 Candidates, and we were down to 133 survivors sitting in the class that afternoon. As the Cadre started slowly calling numbers, we all gauged the roster number being called. Was that a good guy? Does anybody know him? Do you remember him screwing something up? A strong Candidate got called, followed by a guy that peered poorly. We couldn't determine a pattern. The unknown and the unknowable. Each number was called and the slow scraping of his chair on the linoleum floor was like nails on a chalkboard.

Then they called Mike's roster number. At last. We had cracked the code. Mike was definitely getting selected. Now you started praying that your number would be called next. A few more roster numbers were called and the furtive glances around the room told of our fragile understanding of the situation. We were down to 113 guys in the classroom. A bloodbath. Only twenty guys got selected? Total heartbreak. I remember a lump forming in my throat. I had failed. Mike earned it and I didn't.

Then a Cadre walked to the front of the class and slowly turned to us and announced, "Congratulations, you've been Selected." Bedlam. Cheers and shouts. Hugs and high fives. More than a few dusty eyes and for everyone some relieved confusion. How did I make it but not Mike? What on Earth were the Cadre looking for? We met with Mike a few days later to share a beer and a steak before flying home. He had met with the Cadre for his out-counseling. They told him that he performed well but that during that fateful event where he announced "No, don't worry about it. I'll take the hit," he was assessed. This was the hit. 21 day non-select.

All the rules matter. All the time.

Life Lesson: There is only so much that you can "take the hit" on. Know the limits of your authority. Sometimes being a cowboy gets you shot.

All The Rules Matter

photo credit K. Kassens, USAJFKSWCS

Candidate Archetypes

I think that it is important to understand the broad categories that Candidates often fall into. This is helpful because it often serves as an unintentional heuristic for Cadre. They are human, so they are subject to the eccentricities of all humans. If you present a specific personality to the Cadre, they will likely be influenced by it. Both good and bad. I will provide five archetypes to help shape your understanding of this phenomena, but there are more.

The Supervisor – This Candidate is usually an active-duty NCO, typically an E-6. They have been conditioned, by virtue of their normal job, to supervise. They normally have a squad at their disposal, and they are accustomed to receiving a mission, task organizing their soldiers, and then supervising the completion of that task. This is a completely acceptable SFAS modality, but not all the time. Many of these Candidates struggle to separate their Candidate self from their Staff Sergeant self. On some teams they will find willing acolytes in the younger, less experienced 18X Candidates who have themselves been conditioned to take instructions unquestioningly.

The leaderless training environment demands that someone take charge, but not the same person every time.

And certainly not in lieu of doing any actual work. So, the Supervisor establishes his position early in the day during the apparatus build time and if he never transitions into a doer then he earns himself the title as such. Talker, not a doer. At SFAS, you need to be a doer. PC or Beret.

Map Check Guy- Do not confuse Map *Check* Guy with Map Guy. A good navigator in Team Week is invaluable. The events are carefully planned with little margin for error, so a missed turn can cost a team dearly. A good Map Guy will also keep a team focused with regular callouts of distance and time. If you can keep a team focused on the next 300 meters and a slight bend in the road, then they might not be focused on the remaining insurmountable task of six kilometers and an over-burdened back. This is who you want to be. A virtual lock for top 3 on peer evals. That's Map Guy.

Map *Check* Guy is the opposite. In fact, a persistent Map *Check* Guy can often overwhelm a weaker Map Guy. Map *Check* Guy is usually a physically weaker Candidate and he needs more rest. But he doesn't want anyone to know that he needs a break so instead of just asking to rotate out or calling for a team rest break he just calls out, "Map *check*! Hey, can we get a map *check?*" His hope is that this will force the team to stop, he will get a rest, and nobody will be the wiser. Everyone is the wiser, believe me. Especially the Cadre. What ends up happening is that teams that don't control the Map *Check* Guy take so many unnecessary breaks that they miss their deadlines. Be the Map Guy and keep the Map *Check* Guy in check.

Knot Guy- Sometimes the difference between a little misery and whole bunch of misery is a simple little knot. When you are constructing an apparatus, it is important to task organize appropriately. Don't let the Supervisor get away with too much but make certain that you have the right guys tying the knots. If you can't tie knots, then go away. Nobody will judge you harshly. Go find something else to help with. Leave this task to the pros. A skilled Knots Guy can get the tires bound together for smooth rolling or the poles welded tight to keep the insanely heavy loads from shifting.

The tubular nylon lashings that you will be provided are often old, soiled, and frayed. A good Knot Guy can make them work. A bad Knot Guy either cannot tie them properly or he is very skilled, and he thinks that his only role is to tie, untie, and inspect knots. Be the good kind and get them tied, teach others to tie them, then quickly move on to the next task. Don't rest on your laurels. I once heard a Candidate say that his team should not make him carry too much on a low carry event so they could protect his hands for the precious knots. Like he was some sort of skilled neurosurgeon. He didn't last long.

C'mon Guys Guy- This guy is also known as the Spotlight Ranger. He's always ready with a quick cheer and is eager to be vocal to a level that doesn't match his actual effort. Cadre have an uncanny ability to gauge real effort and I've had many Cadre show me their little tells. The depth of a footprint, the shifting of a load in an explicit way, the tautness of a rope or strap, and even the specific sounds that a ruck strap makes. Save your cheerleading for elsewhere. Your team is brimming with motivation, they don't need that kind of help. Pick up an extra sandbag or do an extra shift under the apparatus if you want to support your team. If you don't, your teammates will be certain to shine the spotlight on you during peer evals.

This Is Awesome Guy -This is who you want to be. Cool, calm and collected. Follows the 3 Rules. Never yells, always listens. When you are being driven into the sands of Camp Mackall under another overweight downed pilot, you glance at him on the other end of the apparatus and he just smiles back and says, "This is awesome." He doesn't yell it, he doesn't piss and moan, and it isn't bravado. You are motivated. He is motivated. There is no need to cheer. He just says, "This is awesome," because it is awesome.

The events are so absurd, and the effort is so immense that it quite literally inspires awe. You will often be rendered speechless. Sometimes you won't be able to say anything because you won't have the breath in your lungs to spare. Other times you won't be able to process the immensity of the task at hand and your words will fail you. In that moment I want you to remember something. You asked to be there. You knew what was waiting for you in the land of the longleaf pine. You knew that this was the barrier to entry, and you needed the challenge. There is no place else that you should want to be. You have a duty. This. Is. Awesome. Be that guy.

THIS IS AWESOME,

PHOTO CREDIT K. KASSENS, USAJFKSWCS

The Long-Range Movement

The final event at SFAS is the Long-Range Movement (LRM). Allow me to get up on my soapbox for a minute and revisit our discussion about the importance of accurate language. Many Candidates refer to the LRM as the Long Walk. This is incorrect. The Long Walk is a different selection to assess into a different organization. When you use the Long Walk to refer to the LRM at SFAS you muddy the water, and you look uninformed. In other words, you violate one of the 3 Rules; you don't look cool. Don't use this terminology unless you are referring to the actual Long Walk. Candidates also refer to the LRM as the Trek. This again is incorrect. The Trek was an exercise that was once part of the 18A pipeline. It had multiple forms in multiple locations and at one point was called the Troy Trek, drawing its name from Troy, North Carolina where early versions of the exercise were centered. There are few, if any, published accounts of this history so let me put my words to the task.

The Trek was a long-distance solo land navigation exercise that was part of the Special Forces Detachment Officers Qualification Course. Its most difficult location was centered in Pisgah National Forest. The terrain is extreme. The Trek was conducted over three days and officer students were tasked with finding points across the forest that when plotted on a map may only represent 35 straight line kilometers of distance, but when translated to actual navigable routes might force the student to move 50 or 60 kilometers. You could travel when you wanted and rest when you chose, but the clock never stopped, and you had to find all your points. Rucksacks weighed in at around

80 pounds. It was a must pass gate and proved to be a formidable challenge, often dropping 20% of a class. I know many excellent Officers who were relieved from the course for failing the Trek and I know one officer who resigned his commission and was readmitted to the Q course as an E-6 18D. He couldn't pass the Trek and he wanted his Green Beret so badly that he made that sacrifice. I think that the Trek was the single hardest event that I ever faced in my career. It was that hard. In the early 2000's it was deemed too detrimental to course output and after a short iteration at the much easier Fort Pickett location, it was removed. But I think that the history of the Troy Trek and the difficulty that it represents deserves to be recognized as its own thing. So please don't call the LRM the Trek. Or the Long Walk.

What the Long-Range Movement represents now is just what the name indicates. It is a final forced road march that stands between being selected and being sent home. The exact distance, time, and route is closely guarded but most sources cite that its about 25 miles and lasts between 6 and 8 hours. I will confirm that the LRM is more than 20 miles but less than 30. Until recently, nobody quit the LRM. In 2021 someone did quit, and it was the first incident that anybody could remember, even the grey beards at SFAS.

Since then, we have seen several more instances of quitting the final event. Some quit within a few miles of the finish line. Unbelievable. Every important decision that you make at SFAS you will make with a ruck on your back.

That is Special Forces Assessment and Selection. Eight times a year Camp Mackall hosts this nearly indescribable event that acts as the gateway to the Brotherhood. Those that attend, even unsuccessfully, cite SFAS as life changing. While the SFAS that you might attend may have a few different events, functionally it will be the same. It will have some cognitive tests and some rucking. It will most definitely have some land navigation. Of course, The Sandman never sleeps. Know that you will be tested. The standards will not be lowered, and you will be treated fairly. Performance is the only thing that matters. This is SFAS.

No Shit, There I Was...The Magic Knife

You may recall Rob from the Special Reconnaissance war story. Well, anyone who is that prone to hysterics is bound to make the stage a few more times. Rob, who was a Ranger qualified Infantry officer, was not very comfortable in the woods. During our Trek exercise he showed us why. Now the Trek was in the Pisgah National Forest which is firmly in the Appalachians. The Appalachians have a well-deserved reputation of being a little creepy if not downright supernatural. Hysterical man + supernatural energy = comedy gold. To add a little heat to the fire, Rob was convinced that we would be Trekking in the height of bear season. To be clear, it was not bear season, Rob did not confirm that it was or wasn't bear season, and bear attacks are incredibly rare.

Rob came prepared regardless. In an exercise where managing the weight of your rucksack became a competition for lightness, Rob spared no expense...in the other direction. He carried two extra-large canisters of bear repellent, a bear alarm, a bear bell around his neck and the most bizarre knife imaginable. He had purchased the largest, gaudiest, cringiest knife conceivable. Some 14-

inch gas station monstrosity of silver inlay, black epoxy, and fake mother of pearl beast with an elaborate hilt and an enormous, curved blade. Something right out of a fantasy novel or a Dungeons and Dragons convention. And it weighed almost 5 pounds. Fully kitted out he looked like a post-apocalyptic prospector.

The Trek was incredibly challenging, and you conducted it essentially unsupported. You could get water resupply when you checked in at one of your points, but you were otherwise left to fend for yourself. Every evening you had to pause and monitor your radio to make a check-in call. Most guys would stop to sleep every night, so it was a welcome bit of entertainment to listen in to your buddies report in with a position update and get a weather forecast while you built your camp for the night.

On the first night out when it got to Rob's turn to check in there was silence. As you can imagine it was a big deal to miss a radio check in, so we listened intently. The extreme terrain and thick vegetation could mask radio signals well, so we didn't freak out too much, but certainly everyone had stopped their fire-tending and meal prep and listened in to see what was up.

"Romeo 1, this is Base Camp, Over."

Silence...

"Romeo 1, this is Base Camp, Over."

Silence...

"Romeo 1, this is Base Camp. If you can hear us just break squelch so we know you're up."

...break squelch...

Interesting.

"Romeo 1, this is Base Camp. If you can hear us break squelch twice to confirm"

...break squelch...break squelch...

Very Interesting.

"Romeo 1, this is Base Camp. If you need assistance break squelch 3 times."

...break squelch...break squelch...break squelch

Oh shit.

"Romeo 1, this is Base Camp. We are sending help to your area, prepare your emergency flare and stand by."

A few minutes pass and we wait with bated breath. The entire class is certainly listening. I am on the edge of my seat, or log as it was.

"Romeo 1, this is Base Camp. We are vectoring in on you. Stand by."
…break squelch…*in a stage whisper* …"BEAR!"…break squelch…

"Last station, this is Base Camp. Say again last transmission"
…break squelch…*in a panicked stage whisper* …"BEAR! BEAR!"…break squelch…

"Romeo 1, this is Base Camp. We copy…bear? Is that correct?"
…louder, but still a panicked stage whisper…"Uhmmm…there is a bear stalking me!"

"Romeo 1, this is Base Camp. Roger that. Do you have your bear alarm?"

…"Uh, bear alarm?...No, I'm separated from my ruck. I made camp and heard a bear and I got up and ran!"

We can all hear across the radio Rob crashing through the brush as he makes his way in the night. A bold move. Stupid, but bold.

"Romeo 1, this is Base Camp. If you're moving without your gear, then stop"
…"Uhmmm…Okay. I'm stopped, But I can hear the bear by my camp!"

"Roger that, stand by. We have help on the way"
…"OH SHIT…I CAN SEE SOMETHING!"

"Romeo 1, this is Base Camp. Calm Down, we have help on the way. Hang on!"
…"OH SHIT….IT'S FOLLOWING ME. THE BEAR, IT'S FOLLOWING ME!!!"

"Romeo 1, this is Base Camp. Does the bear have a flashlight?"

…

...

..."Uhmmm, yeah I see a light."

Then, a new voice comes on the net. The senior Cadre in an exasperated voice, "Bears don't carry flashlights, that's me you dickhead. Will you please stop running so I can check up on you?!"

Then, as if on cue, another student breaks the silence with a clearly disguised voice.

..."Hey Romeo 1, why didn't you just use your Magic Bear Knife...you dickhead!"

Always look cool.

Life Lesson: Less superstition and more analysis. You don't have to be a smart person; you just have to think like a smart person.

Chapter 4 - How to Prepare for SFAS – Gate Week

Chapter 3 provided us a good summary of what SFAS is in terms of both scope and sequence. Over time some of the individual tests may vary but they will measure the same general attributes. Some events may get rotated for new events, but they will have the same overall characteristics. You will have to be smart. You must ruck well, you need to be able to navigate, and you must be physically fit. You need to be a skilled communicator and learn to listen well. You should be a critical and creative thinker and thrive in ambiguity. A question that often gets asked is, "What are the Cadre looking for?" This is a difficult question to answer because the answer is that the Cadre are looking at everything and they are looking for everything. Broadly speaking the Cadre are looking for good people. Good people make good citizens, good citizens make good soldiers, and good soldiers make good Green Berets. Officially, the SFAS Cadre mandate is to select the Candidates that can successfully finish the Special Forces Qualification Course. In practice, they are not looking for just the best people, they are looking for the right people. They are looking for people with the Army Special Operations Forces (ARSOF) Attributes.

Since its inception in 1988, SFAS has sought to quantify what specific attributes that it was assessing and selecting for. Originally it was a fairly nebulous set of *SF Core Values* that remained officially undefined, but included good judgment, self-discipline, responsibility, independence of mind, stamina, patience, a sense of humor, spiritual toughness, and maturity (Duffy, 1999). There was never a definitive set of behaviors that was associated with these values, so assessing them was always an exercise in, "I'll know it when I see it." In the late 1990s the focus became the *Army Values*, which were well defined but again lacked any real associated behavior to aid Cadre in assessing Candidates.

In the early 2000's, SWCS started to develop its own unique attributes to help screen Candidates. The initial research was focused on the concept of *Adaptability*. The initial definitions were founded on the I-Adapt Theory (Burke, Peirce, & Salas, 2006) and further refined by SWCS (Mueller-Hanson, Wisecarver, Dorsey, Ferro, & Mendini, 2009). SWCS used the below framework for initial analysis.

Adaptability	
Physically Oriented Adaptability	**Environmental Resolve**: adjusting to environmental states; **Physical resolve**: pushing one's self to complete physical tasks, or **Physical Fitness**
Learning Tasks	**Proactively:** searching out learning opportunities in advance of a challenge or **Reactively:** doing what is necessary to keep up; **Self-Reflection**: taking action through the process of reflection
Handling Emergencies or Crises	**Responding to Life-Threatening Emergencies** (i.e. sniper); **Administrating First Aid--** applying medical techniques
Handling Stress	**Resilience & Emotional Control**: remaining calm & in control of emotions;

	Extraordinary Stressors: under fire, extreme noise, negative feedback, etc.
Dealing with Change or Ambiguity	**Cognitive Flexibility:** responding to changing environments; adjusting one's perspective; **Flexible Decision-Making:** Changing approach to a problem; **Critical Thinking:** Identifying powerbrokers, motives, leverage points; situational awareness
Thinking Creatively	**Problem-Solving:** developing innovative solutions; **Working with Limited Resources:** methods of obtaining or utilizing resources; **Information Gathering:** using innovative methods to

	understand links and capitalize on opportunities
Cultural Adaptability	**Negotiation of Cultural/Language Barrier**: communicating effectively across cultures; **Cultural Tolerance:** willingness to adjust behavior or show respect to others' customs or values
Interpersonal Adaptability	**Building Rapport Influencing the Dilemma/Negotiation:** communicating with awareness of leverages; **Membership & Participation as a Team Member:** working with others

But *Adaptability* did not cover all of the attributes that Special Forces leadership thought encompassed all that a Green Beret should be. SWCS Commanders spent about a year querying the force, specifically the USASOC

Commanders Conference, to help develop a more comprehensive list. A full list a of about 50 attributes was circulated to Senior Commanders and Senior Enlisted Leaders and they were asked to rank order them. The list was re-prioritized, condensed, and recirculated for comments until the final list of ARSOF Attributes was created.

In 2010 SWCS finally codified the ARSOF Attributes. Most importantly SWCS also developed a guideline for Cadre that included an exhaustive list of real behaviors, aligned with specific selection events, that were indicative of the ARSOF Attributes. The military loves a checklist and SWCS is no exception. Again, academia provides us some real insight into this list. In one study, the original publication of the ARSOF attributes provides us some insight into these behaviors.

These attributes with associated components and reconciled definitions are listed in the below table (Richmond & Roukema, 2010).

Main Standard or Expectation	Components	Reconciled Definition

Courage	Displays: Moral Courage Physical Courage	Acts on own convictions despite consequences; is willing to sacrifice for a larger cause; not paralyzed by fear or failure.
Integrity	Is: Trustworthy Honest Ethical Loyal	Being trustworthy and honest; acting with honor and unwavering adherence to ethical standards
Perseverance	Is: Motivated Committed to Outcome Mentally and Physically Tough	Works towards an end; has commitment; physical or mental resolve; motivated; gives effort to the cause; does not quit
Personal Responsibility	Is: Self-Disciplined	Is self-motivated and an

	Accountable Autonomous Motivated Displays: Initiative	autonomous, self-starter; can anticipate tasks and acts accordingly; takes accountability for his actions
Professionalism	Displays: Maturity Humility Judgment Decisiveness Confidence Self-Discipline Initiative Self-Awareness Is: Articulate Emotionally Stable	Is a standard-bearer for the Green Beret; has a professional image to include a level of maturity and judgment mixed with confidence and humility; forms sound opinions and makes own and stands behind his sensible decisions based on his experiences

Adaptability	Is: Innovative Resilient Culturally Astute Influential Aware Self-Reflective A Critical Thinker A Problem-Solver A Flexible Thinker	The ability to maintain composure while responding to or adjusting one's own thinking and actions to fit a changing environment; the ability to think and solve problems in unconventional ways; the ability to recognize, understand, and navigate within multiple social networks; the ability to proactively shape the environment or circumstances in anticipation of desired outcomes

Team Player	Is: Dependable Selfless Respectful Loyal Committed to team/mission	Able to work on a team for a greater purpose than himself; be dependable and loyal; work selflessly with a sense of duty; respects others and recognizes diversity
Capability	Is: Physically Fit Technically/ Tactically Proficient Intellectual	Has physical fitness to include strength and agility; has operational knowledge, and ability to plan and communicate effectively

The actual checklists, produced in various formats including pocket guides for Cadre, are confidential. But they represent a real effort to quantify what is looked for when assessing Candidates. Newly assigned Cadre go through a rigorous onboarding and training process so they can standardize expectations and ensure fair and impartial assessments no matter which Cadre is making the observations. So, if you want to know the answer to the question of "What exactly are the Cadre looking for?", now you have the answer. A long explanation of how we got the list, a few charts to describe the list in detail, and specific components of the list. The question now becomes how you prepare yourself so that you can be assessed. How do you get selected? For younger guys, I have included an entire chapter just for you. For hopeful Candidates that are ready for the commitment, this next part is for you.

Gate Week Prep

The PFA

The Physical Fitness Assessment is the first assessed event at SFAS, and it is the only event with publicly published standards. The standards are surprisingly low, and the most failed event (sit-ups) has been removed. You also are required to have successfully completed a PFA within 30 days of reporting to Camp Mackall. Despite this seemingly modest and obvious task, we still see classes with as many as 20% drop rates for the PFA. A fair percentage of Candidates pass the PFA but score so low that they are statistically pre-ordained to not complete SFAS. I have personally observed dozens of PFAs and I think they have all been some of the easiest and most fairly graded PT tests in the Army. Gone are the days of Cadre no-counting repetitions to winnow student loads and the Cadre make great efforts to ensure that they are all grading to the exact same standard across the board. If you can't pass the PFA at SFAS then I would really question your commitment to the process. If you can't pass the PFA convincingly, then you are in for a long three weeks, likely much shorter.

Many Candidates in training fall down the rabbit hole of what I call *program jumping*. That is, they want to prepare adequately but they lack a real plan. There is no shortage of plans from self-branded SpecOps/SFAS/Tactical Athlete/Warrior/Flaming Skull fitness experts, but most are incorrectly focused. I have seen many of these experts cite their fitness credentials and they are impressive, but most have no relevant SFAS knowledge. SFAS is a unique beast so you must prepare in a specific manner. Experience and expertise matters. Many Candidates in training are lured by the slick marketing and promises of selection success, but they lack the discipline and direction to follow through. As a result, they end up chasing promises and program jumping. A side-effect of this is that they lose the focus on PFA events. I believe this accounted for the high attrition rates from sit-ups. Most fitness programs don't train sit-ups, they train core strength and stability. The unfocused and undisciplined Candidate in training ends up missing a key component.

What does this foretell for the rest of SFAS, where they don't publish the standards and the events remain unknown? Pick your expert advice wisely. Keep your attention on what the current PFA entails. The Special Operations Recruiters will be your most valuable source of current information and the National Guard Training Detachments are well-informed. Seek out their official social media accounts. Then ensure that you incorporate a diagnostic PFA into your training regimen at least once a month to ensure that you are training appropriately. Adjust as needed.

Cognitive Tests, CRA, and Nasty Nick

The cognitive tests are much more difficult to 'train' for. In certain respects, you are either smart or you are not. You can't really study for an IQ test. They do publish prep programs for some of the standardized batteries including some of the ones at SFAS. But the efficacy of these prep programs is unproven.

The Combat Readiness Assessment is about an hour long and there seems to be plenty of rest between events, but it is still a grinder. It requires strength, speed, coordination, and no small amount of perseverance. You should train athletic movements and mobility. Compound lifting movements, kettlebell work, asymmetrical strength work, and explosive movements. I don't recommend training for the CRA separately, rather incorporate these elements into your overall training plan. In a word, athletic. Faster than the strongest lifter, stronger than the fastest runner.

Prep for the Nasty Nick should be an extension of your prep for the CRA, with the inclusion of rope work. Climbing a rope definitely requires a strength component, but good technique is really the key. There are several obstacles that incorporate rope climbing and they are spread throughout the course, so by the time a Candidate reaches the closing obstacles strength and grip are significantly diminished. Add to this the fact that Candidates are also likely wet and muddy, then relying on strength simply isn't tenable. I would also spend some time working the transition mantle at the top of a rope climb.

Most military rope climbs require the climber to just touch the near pinnacle of the rope. The Nasty Nick will require you to climb all the way up and mantle over the top of the anchor point.

ROPE CLIMBING IS ONE THING,
MANTLING OVER THE TOP IS ANOTHER
PHOTO CREDIT K. KASSENS, USAJFKSWCS

The Nasty Nick will also test your tolerance of heights and enclosed spaces. There are multiple obstacles that require Candidates to climb high, balance at height, and negotiate tunnels with limited visibility. It is not uncommon for Candidates to freeze at height and panic in the dark. Assessment is ongoing. If you have access to a standard military obstacle course, then you should practice often. If you do not have such access, then I would recommend the closest facsimile possible. Many communities have fitness trails that may produce some of the same outcomes, but at a certain point we need to adhere to the specificity principle of our fitness planning. Work on grip strength as this is critical for Team Week as well.

No Shit, There I Was...Pig Pen

A frequent question that I get asked is about the Psych Eval. "What are they looking for? Will I get dropped for this thing that happened?" The short answer is, "it depends." The long answer is that it doesn't matter. You can't really study for the Psych Eval, and you should not be looking for ways to game the system. You might be able to manipulate your answers around a topic sufficiently, but SFAS is designed to expose all your behavioral anomalies. Candidates are uncovered quickly. In my class we had Pig Pen. He was easy to figure out.

Pig Pen was a young Military Policeman from Fort Bragg. He attended SFAS to get out of a JRTC rotation. This tactic of attending SFAS to avoid a CTC rotation was fairly routinely used, especially for Candidates assigned to Fort Bragg. It was so common that SFAS had a policy that you couldn't quit until Day 7. I recall that I had fire guard on Day 6 at midnight. When the clock hit midnight a bevy of alarms chimed and about a half-dozen guys got up, fully clothed, and dragged their duffel bags off to the Cadre hut to quit. But not Pig Pen. He was a bit different.

Pig Pen started his SFAS journey fully intending to quit. But when he showed up to Camp Mackall he found that he wanted to give it a try. The only problem was that he intentionally flubbed his Psych Eval to guarantee that he wouldn't have to continue training. But he wouldn't quit. He got a 21-day non-select. They offered him an immediate return. He took it. He went through SFAS again. 21-day non-select, again. Two consecutive 21-day non-selects. They offered him another immediate return. He took it. When he reported to my class it was his third consecutive class.

Suffice it to say, Pig Pen was *rode hard and put away wet* by the time he was classed up into my class. Walking wounded. Bruised and beaten, and a little bit broken. Upon arrival, Pig Pen just decided to stop doing personal hygiene. His uniforms and equipment looked like a guy who just done SFAS twice in a row. He grew a nice beard and had a perpetual cloud of funk that followed him around wherever he went. His feet looked like raw hamburger. I can't imagine doing SFAS twice. I can't fathom doing it twice in a row. Pig Pen showed up three times.

I'm not sure what lesson we can learn from Pig Pen, but I will never doubt that kid's heart. His head? That's another story.

Life Lesson: Don't self-sabotage. It's easy to be your own worst enemy, don't make your life harder than it has to be, never self-select.

THIS ISN'T PIGPEN, BUT I'LL BET THAT HE SMELLS LIKE A SWAMP

PHOTO CREDIT K. KASSENS, USAJFKSWCS

Running Prep

Your running prep should be an extension of your rucking prep in that your running is a component of your total weekly mileage. If you follow the rucking guidance of shorter more intense sessions, usually 2-3 times a week, (see rucking prep section for this reveal) then your run programming should be complementing that effort. You must build aerobic capacity. Some Candidates in training can tolerate higher mileage. As your prep programming advances and your rucking efforts intensify, you should appropriately taper your running. I don't think there is a magical number of miles as athletes respond differently to similar stimuli. Speed work, including sprints, are a valuable component. Slower distance runs are helpful as well.

One component of running prep that most Candidates get wrong is intensity. The conventional wisdom is that if you want to run faster, then you must train faster. The data is clear that 80% of your running should be done in the Zone 2 spectrum. Zone 2 effort is 60-70% of your maximum heart rate. These runs should be limited to 90 minutes or less. As you build your prep program you should start with more running than rucking. As you build strength, and you take on more rucking stress you should have built a strong aerobic base and you can taper your running mileage. This is a good opportunity to incorporate speed workouts. Monitor your training carefully and seek to balance effective ruck training, effective run training, and manage overall mileage to prevent injury.

Effective strength training can also help your running, particularly with common running injuries. Achilles strains, shin splints, and plantar fasciitis can be mitigated with proper, targeted strength training. These issues, as well as stress fractures, are also common in overtraining. When Candidates in training fail in their prep it is most often in the timing. They try to cram six months of training into eight weeks. You must build your strength, then build your aerobic base, and then build your speed. You must be deliberate in your run programming.

The answer to the question of how much running and what type is, "It depends." It depends on your individual tolerance for mileage, it depends on running form and efficiency, and it depends on your rucking performance. One of the biggest mistakes that Candidates in training make is overtraining. Going to SFAS injured, even a minor injury, is a recipe for disaster. You will not have time for any recovery and even small injuries have a way of cascading into total failure. A little limp from a sore ankle can alter your gait only slightly, but when your mileage starts to add up that little limp starts to cause issues in your hip and then your back. What started as a nuisance is now a real show-stopper. Learn to recognize the difference between being hurt and being injured and learn to temper your enthusiasm. Rest when it is warranted and allow your injuries to heal. This is why a good prep program has some inherent flexibility to allow for minor setbacks and slower or faster adaptations. Train hard, but also train smart.

A*T *SFAS*, R*UCKING* I*S* K*ING

PHOTO CREDIT K. KASSENS, USAJFKSWCS

Rucking Prep

You can't call your book *Ruck Up or Shut Up* and not give significant focus to rucking. You can't go to SFAS and expect to be successful and not give significant focus to rucking. So, we will dedicate about 7,000 words or more than 10% of this book to this specific topic. I have exhaustively researched this and have compiled the most relevant literature to support my conclusions. This analysis isn't opinion, it isn't based on anecdotal narratives, and it isn't inconclusive. You may know a buddy who didn't ruck much to prep for SFAS, and he made it. You may have heard about time under load or long slow distance techniques from an article you read on the internet. You may have seen tips about inside out socks and body glide from some guy in your unit. Leave that stuff in the periphery. You need to focus on evidence, not narrative. You need to focus on proven, empirical, and pragmatic techniques. This is what I will provide.

Rucking is the most important performance predictor for SFAS success. Rucking is king. Every important decision that you make at Selection you will make with a ruck on your back. There is a reason why I will repeat this often and continue to reinforce its importance. The evidence is conclusive. You must learn to ruck well, repeatedly, and continuously. You must learn to think well with a ruck on your back. Most ruckers just throw on the ruck and get into a zone where they specifically don't think.

SFAS Candidates do not have that luxury. Rucking is part fitness, part technique, and part misery management. An art and a science. This prep will address all of these elements.

There seems to exist a corps of naysayers that believe that rucking is inherently bad, that it produces too many injuries. There is not much evidence to support this. I'm relying on my experience and the lack of evidence supporting this conclusion. But rucking, when properly programmed, is not dangerous and does not produce higher injury rates. I think that rucking too hard, too soon, or too heavy is definitely bad. Most military units do rucking so infrequently that when they do finally execute the task, it becomes this emotionally significant event. You hear all sorts of urban legends and old wive's tales about nylon dress socks, Preparation H, and putting your socks on inside out. This is nonsense and anyone who gives you this advice can immediately be dismissed as an amateur. Ignore them. We will cover foot prep later in this chapter.

I will organize the rucking prep section to address the topic in a logical and comprehensive manner:

<u>Rucking Fitness</u>- building prerequisite strength; programming methodology (frequency, intensity, conditions)

<u>Rucking Technique</u>- understanding pace; learning to shuffle

<u>Misery Management</u>- handling wounds; managing cognitive load; building the right musculature

<u>Foot Prep and Care</u>- boots, socks, and feet; pre-SFAS blister care, SFAS injury treatment

Rucking Fitness: Building up rucking fitness takes time. You need to build a requisite baseline of overall strength first. Until you reach this baseline strength level your rucking performance will be stifled. You need to build up your weight tolerance slowly. Too much weight or weight too soon will inevitably lead to injury and setback. Never add more than 10% weight increase from week to week, perhaps a little more when the weight is very low. I will draw on two studies that seem the most applicable and are very strongly reinforced by my experience. Mountain Tactical Institute (MTI) sponsored a series of studies that demonstrated the baseline strength and frequency requirements that I cite. The MTI studies establish the baseline strength requirement in Study #1 and the methodology in Study #2. Critics might question the study population, relatively unconditioned cadets, as being a poor likeness of the Candidate in training. I think this is an excellent representation as the age and fitness levels are very similar. Most Candidates in training will have little prior exposure to rucking.

What the MTI studies demonstrate is that building strength, as measured by Front Squat Relative Strength and Bench Press levels establishes the baseline abilities to allow subjects to then progress appropriately in the rucking specific adaptation (Scott, Ruck Deep Dive – Study #1: Physical Attributes Which Relate to Rucking, 2015). Once that baseline fitness level is reached, the conclusion is that field-based progressive load carriage, usually 2-3 times a week is optimal to build rucking performance (Scott, Shaul, & McCue, Ruck Deep Dive: Study #2 - Ruck Training Adaptation, 2015). Let's decipher that analysis.

The baseline strength requirement as measured by Front Squat Relative Strength and Bench Press states that you should be able to squat ~1.25 x your body weight and bench press ~1 x your body weight. The calculation is not assessed as causation, but it is very clearly correlated that the stronger your squat and bench press the better your rucking performance is. The MTI studies also draw on significant literature that supports this conclusion. Build strength first, then start rucking. This is counter to many prep programs that start with a focus on rucking before significant strength gains are established. But the preponderance of the literature is very clear, build strength first, then start rucking. Next, let's decipher the methodology statement. Field-based progressive load carriage, usually 2-3 times a week.

Field-based refers to the specific training environment. You should endeavor to replicate Camp Mackall as much as possible. If you can find sand, packed and loose (the looser the better), then you will be well-served. There isn't a great deal of elevation at Camp Mackall, but even a small incline will present a significant challenge. If you train primarily on asphalt or a performance track, or heaven forbid a treadmill, then you will be poorly conditioned for the reality of the Sandhills. Look for gently sloping offroad routes with packed and loose sand.

Progressive load carriage refers to the methodical increase to both distance and weight. Start light and short. Just a few miles with 25 pounds is fine. Work with these parameters for a few weeks. You will likely adapt quickly and feel the urge to go longer and heavier. Fight that urge early on. You are adapting in multiple ways during these early sessions. Not only are you building fitness, but you are also building resilience. Resilience in your feet, actually toughening the skin, but mental resilience as well.

Remember, rucking is fitness, technique, and misery management. It's not very miserable yet, but the threshold for misery is easily reached.

The final element of the MTI studies is the frequency, *usually 2-3 times a week.* You will see many SFAS prep programs incorporate only one ruck a week, or even skip consecutive weeks. Remember the fitness principles: specificity, progressive overload, and adaptation. Rucking is a fitness event at this stage of preparation so we should apply the principles as such. The flexibility of 2 or 3 rucking sessions per week is based on a couple of factors. Your total mileage that week, your persistent recovery, and the weight you are carrying impact the decision to incorporate another ruck or an active recovery session. Don't overthink your ruck programming. Keep it simple. Not easy, but simple: Field-based progressive load carriage, usually 2-3 times a week. Start slow and low, increase distance and weight as you adapt. Simple.

Rucking Technique: But what about speed? We could also measure speed as intensity. Here we have another study to help base our preparation. A 2014 study indicates that multiple shorter more intense rucks are superior to longer slower rucks (Maladouangdock, 2014). Again, this study is well reinforced by supporting literature. This invalidates the conventional wisdom of most of the available fitness programs. If we synthesize the MTI and Maladouangdock studies what we get is **field-based progressive load carriage, usually 2-3 times a week focused on short intense sessions**. That is the recipe for programming ruck workouts. If your fitness guru recommends something different, then you should ask why. What evidence is used as empirical justification? If the ruck programming violates this, then what else is wrong? Rucking performance is the most important, so if the big stuff isn't right, how good can the little stuff be? Time under load is a common fitness programming discussion, but it does not apply to rucking. Field-based progressive load carriage, usually 2-3 times a week focused on short intense sessions. That's it.

The time, distance, and weight standards for the ruck events at Selection remain confidential. What we can surmise is that the Army standard of 15-minute miles with 35 pounds is not very rigorous and training with those goals is not likely to result in getting selected. I usually recommend the training goal of 12-13 minute miles with about 55 pounds of weight. If you can consistently produce these results, then you should be well-prepared. You can expect to be fatigued and for conditions to be very challenging. Given this time and this weight, you should start to focus on building your speed.

There are only two ways to increase your speed. Either increase your pace or increase your stride length. Lengthening your stride is tricky, so proceed with caution. You might end up altering your gait so much that you produce unintended consequences and injuries in your joints. It is easy to tweak your knee or jam up your hip. An excellent training technique is to have a coach or training partner video your movement under different conditions. Use that video to analyze your gait and look for erratic movement and seek to develop fluid and consistent motion. Slow is smooth, smooth is fast. The other method to increase your speed is to increase your pace. Simply moving faster isn't always so simple and you should be mindful that small changes can build big results in the aggregate.

The standard velocity is about 60 paces per minute. A "double-time' pace is obviously twice that, at 120 paces per minute. This is a running pace. The goal then is to train to match your pace and standard stride length to result in 12-13 minute miles. That's a fairly straightforward proposition and with a stopwatch and a GPS or pedometer you can fine tune your personal speed. The goal should be to not run. Running with a ruck produces some tremendous stresses on your joints, particularly your back and knees. Any awkwardness to your gait can quickly produce suboptimal ankle, foot, and hip movements as well. You should definitely avoid running until you have built up considerable musculature to support this movement. A helpful training aid in maintaining a specific pace is music. You can match the beats per minute to regulate your pace and there are multiple music services that allow you to filter for this feature. A few extra steps per minute can yield significant results as the mileage accumulates.

To understand the impact that just a few steps per minute faster pace can make, you can perform some simple math functions. At a pace of 100 meters per minute you are performing a 16:06 mile. At a pace of just 110 meters per minute that time drops almost 90 seconds to 14:38 a mile. 15-minute miles is the widely accepted Army standard. Over the course of a 10-mile movement the difference is a staggering 14 minutes and 40 seconds. That's just a meager 10 meters per minute increase. Little more than a few shuffling paces can make a massive impact in the long term.

The Shuffle

If you can't walk fast enough and you want to avoid full-on running, then you need to develop the shuffle. You will recall that I said that rucking was part art and part science. The science part is the pacing calculations, kinesthetic movement analysis, and progressive load carriage programming. The art stuff is learning to develop a shuffle. A shuffle is not a full-on run, where the foot strike occurs in front of the athlete, or a walk where the foot strike occurs directly below the athlete. A characteristic of the run is also a continuously bent leg, as opposed to a locked leg on a walk. So, for endurance, a walk is preferred in that the leg relaxes, even for a very brief moment, with every step.

A shuffle seeks to maximize the increased pace of the run but minimize the increased impact. In an attempt to get the best of both techniques you need to spend some time figuring out what your shuffle looks like. I used to train with a fellow Green Beret whose shuffle included this very violent leg whip when he extended his leg forward. It was so violent that he once broke his foot during a ruck due to the motion. But he was fast. I would try to emulate it and it just wasn't for me. The takeaway is that you need to work on what technique produces the best results for you. The balance of speed, endurance, and injury prevention is difficult to figure out, but once you do, you will have a powerful and sustainable tool for improved rucking performance.

Misery Management: I think that we've well-covered the fitness parts and technique parts. Let's move on to the misery management part of rucking. More art and science. Rucking is a unique fitness event. It simultaneously stresses your cardiovascular system and total body strength. SFAS rucking requires high V02 max capacity while also requiring significant muscular strength and endurance. Not just force in the traditional sense of mass x distance divided by time, but strength to endure. There is a unique way that the continuous press of the ruck straps cuts into your trapezius that is a distinctive type of misery. If your SFAS fitness prep plan doesn't include some trap building exercises, then your rucking is likely to suffer. It is not uncommon for newer ruckers to experience nerve pain and numbness in the arms and hands from the continuous pressure.

MISERY MANAGEMENT

PHOTO CREDIT K. KASSENS, USAJFKSWCS

It is also not uncommon for Candidates to experience significant abrasions on their shoulder, hip, and back. Bloody shirts and pants are not unusual and if left unattended these wounds will quickly become infected. Even minor skin injuries are cause for concern. Cellulitis is the inevitable conclusion, especially given the embattled immune system and less than perfectly hygienic environs. Personal hygiene and cleanliness become critical selection skills, not just virtuous personal habits. I'm not certain that building callouses on your hips is a viable option but learning to deal with the misery of broken skin and how to properly manage these wounds for long term austere care will certainly serve you well. We will cover foot care, to include blister management, later in this chapter.

What we end up with is this nexus of rucking anguish created by the continuous and complete fitness burden, the alarming physical pain element and this rising cognitive load. There is just something unique about the way the mind starts to insulate itself when the ruck continuously squeezes into the shoulders and back. It is a natural response and if your task is to simply conduct a roadmarch then there isn't really an issue with this. You can shut down your mind and drone out without issue. But SFAS is different. The majority of what you do is done with a ruck on. You are continuously assessed across multiple events and multiple domains while wearing a ruck. If you want to engage in those assessments in a diminished mental state, then you will likely not assess well.

Land navigation at SFAS is an excellent example of this phenomena. Land nav is a skill. It is also a particular type of cognitive load. You must think deeply when plotting points on a map and calculating and converting azimuths and distances. You must think critically when determining routes that take advantage of terrain and avoid obstacles. You must be disciplined in your time management and remain mindful of rules and restrictions. There is a significant cerebral burden. At SFAS, unlike most land navigation exercises, you must do this with a ruck on your back. Candidates must manage all of that misery that the ruck brings while keeping a clear head. If your SFAS prep only includes mindless rucking, then you will likely not be successful. Throw in some cognitive tasks along with your rucking workouts. Every hour perform a word scramble or similar task. Learn to engage your brain effectively while under load. Learn to manage your misery.

The Ruck
The Tick. My Friend Alice. The Green Heater. The Ruck-a-sac. Call it whatever you want, this is your new best friend. You must learn to become one with your ruck. Don't develop an adversarial relationship with your ruck, you're about to have a lifelong connection. Just embrace the suck. The sooner you start to learn how to manage your ruck…pack it, put it on, adjust it…the better. As of the Summer of 2023, the MOLLE II is the rucksack used at SFAS. You should endeavor to train with this specific ruck. You could train with any number of packs, but you would be well-served to start to understand the nuances of this particular ruck. A weighted vest is an inferior facsimile as it carries the load differently than a ruck. You get the resistance, but not the specific resistance of a ruck, so I would avoid a vest.

The best combat rucksack ever is the ALICE ruck (hence the nickname), and I would encourage you to adopt this superior pack as soon as your circumstances allow. I would also encourage you to read *The Baldwin Articles – ALICE Pack Trilogy* by LTC Terry Baldwin, US Army (Ret). It is an exhaustively researched and very well-written history of the ruck and he does a masterful job of explaining why it is the best combat ruck. It is the definitive work on this topic as far as I am concerned. I recently acquired an early generation lightweight rucksack from circa 1963 and I have adapted this unique frame to my modified ALICE pack. In doing so, I have discovered nirvana. I don't ruck with the intensity required for SFAS anymore, but I look forward to many years of use.

The proper way to 'backpack' is to use the hip belt to take much of the load off of your shoulders and carry it on your hips. All the issued rucks come equipped with a hip belt to facilitate this. But that's backpacking, and we are talking about rucking. Rucking is different. Eventually you will wear your ruck in a tactical environment. In this environment you will often have to react to contact or otherwise shed your ruck quickly. The shoulder straps are equipped with quick-release buckles to enable this action. Even absent of the duress of chaotic combat conditions, I have seen countless hapless Candidates and beyond get caught with a heavy ruck tethered to their waist. The hip belt buckle is not designed to release under pressure, and this is a difficult position to recover from. And it definitely doesn't look cool. I strongly recommend training without the hip belt.

The SFAS Packing List does not allow any modified shoulder straps or hip belts, but when you first start rucking you might consider some additional padding on your shoulder straps. Many novice ruckers experience nerve pain and numbness in their hands and arms because they haven't adapted to the straps yet. So, I would avoid the hip belt, but I see some merit in extra padding on the shoulder straps, especially early in your adaptation. Little tweaks to your shoulder straps can yield big changes in how it feels. At a certain point, it just hurts, but you should learn to manage the misery. The sternum strap is helpful in fine-tuning strap placement but make certain that it is positioned so as not to interfere with the quick-release buckles or that it doesn't stay fastened and guillotine you when you activate the quick release buckles.

When you don your ruck, I would recommend always doing it a particular way. This method won't be possible when your ruck is *really* heavy (in excess of 100 pounds), but those times are few and far between and I would strongly recommend against training with excessive weight. For those infrequent times when weight is too excessive you really only have one option for donning your ruck. Sitting down, sliding into the straps, and buddy-assist standing up is about your only option. For the rest of the time, I recommend the method of standing with the ruck at your feet with the frame facing away, grabbing the ruck by the frame, lifting it upside down over your head, and sliding into the straps. What this method does is it allows you to reveal any packing deficiencies. If you were digging in your ruck and for some reason you forgot to secure it, this method would allow the flaps to drape down or loose items to drop. It's a simple method that ensures your stuff is always correctly packed and secured. This is especially helpful during land nav and if you correctly develop strong RV procedures, you will never lose gear. Even when your cognitive load is being taxed.

The final ruck component is packing. When you pack the ruck, the heaviest items should be placed high on the back and closer to your body. Avoid using water as your weight for rucking. Unless it is very well packaged the water tends to shift around and create an unstable load. Commercial ruck plates are excellent but can be very expensive. You can create your own ruck plates by affixing weights to a piece of plywood cut to shape. I have several plates that I made with mismatched garage sale weights that cost me less than five dollars total, contrasted with a hundred dollars or more for name brand commercial variants. You can also get a fifty pound bag of play sand from a big box store for around five dollars. Just make certain that you reinforce the plastic bag, or you'll be cleaning sand constantly.

    Common sustainment items should be packed with ease of access in mind. Socks and foot care items should be near the top of the pack or placed in outboard sustainment pouches. Use similar procedures for food, water, and cold weather gear. Low use items, like extra boots, uniforms, and sleep systems, should be stored at the bottom. You should learn to locate critical common use items in the dark. Your ruck will become your mobile house, so learning to access all of your stuff without a ruck sack explosion, will serve you well. Guys build oddly strong bonds with their rucks. Near endless accessories, pockets, and custom straps help you to make your mobile house a home. Manage your home well, and you'll live a happy life.

Coaching: This is probably a good place to address coaching. There are so many options available for coaching services and with so many options there is a broad spectrum of quality associated with the coaching, as well as pricing. Expensive coaching is not synonymous with good coaching, nor vice-versa. If you are an experienced athlete and have good awareness of your mechanics, thresholds, and programming principles then you can likely forego these services. But experienced athletes also understand how effective good coaching can be when appropriately applied. If you are less experienced then good coaching can be a critical component of your prep, especially in the early stages of your journey. Explosive and powerful compound lifting movements can be dangerous if you are not following strict form. You might also benefit from very specific coaching services on a limited basis. Having a consult with a performance nutritionist can be immensely helpful. The standard American diet is not supportive of elite physical performance. If you carefully document your training progress, you are likely to see the direct impacts of good nutrition on your performance. So, a few hundred dollars investment early in your prep can give you much bigger returns on investment as your program gains in intensity. The same can be said for running. You might not need to keep a running coach for the duration of your prep, but consulting with a good coach early in the process to help you work on form and mechanics or help develop a good plan with milestones can pay dividends long after the initial sessions.

General fitness coaching is a little more difficult. Concurrent with my earlier comments on program jumping, the fitness industry is just that, an industry. Any industry is going to develop a profit motive and will inevitably attract service providers who seem focused on profit over quality product. Unless the athlete has a specific injury or a medical issue, the programming should be relatively transferable. So, a simple comprehensive program, with some room for individual adaptations, is likely sufficient: Buy the program, include an initial consult, and allow the athlete to go to work.

I'm leery of the online coaching trend as it seems to default to subscription services, and those services seem unreasonably expensive. For most athletes this seems unnecessary. But I also recognize that many Candidates in training, especially younger less experienced ones, need more guidance. They often lack the discipline to stay focused on specific program objectives and motivation seems to always play an important role. For these cases, maybe a few hundred dollars a month of coaching or mentorship is warranted. That accountability might be the missing link.

Candidates in training should be critical of potential coaches claimed qualifications. A Certified Strength and Conditioning Specialist is an obvious credential, but they should have some specialized experience in SFAS. We seem to see quite a few coaches who tout D1 Collegiate experience, but that qualification can vary wildly depending on the program or the individual's role in that program. Proceed with caution. A Green Beret is a logical qualification as well. Who better to coach you than a guy who has been to SFAS? To be blunt, most Green Berets know very little about Selection[9]. They have a singular experience with SFAS and in that experience they were recipients, not observant contributors. Being a gunshot victim is not a qualification for defensive firearms instructors. Being a SEAL is certainly not a qualification for competent SFAS prep advice or coaching. An

---

[9] I'm not throwing shade at my brothers; I'm just noting a reality. I've spent significant time surveying this information environment and the lack of specific SFAS knowledge is shocking. I have subscribed to multiple services with a nom de guerre and I have yet to find one that I would endorse. The information is out there, and now that I have synthesized it in this book, I would hope to see some significant improvements.

immediate red flag is when a program starts making up exercises or renaming existing exercises to meet their 'proprietary' programs. Simple is almost always better. And we should be especially cautious of any claims of extraordinarily high success rates or even anonymous endorsements. SFAS is a very specialized performance environment with many nuanced assessment parameters. The title of former D1 coach, Recon Marine, or CrossFit Franchisee mean virtually nothing at Camp Mackall. Choose wisely.

Foot Prep and Care

In terms of SFAS Prep, foot care may rank amongst the most important. A bad blister during prep can delay your training for weeks. These are valuable weeks that you need to be working. Infantryman and other ground-based combat arms jobs understand that your feet are the foundation of good field living. However, warm, dry, and healthy feet are not the norm at SFAS. We've already discussed the SFAS injury rates and types, but those reports are only based off Candidates who seek medical care. Everyone at SFAS gets blisters, they just don't all seek medical care. Every time this topic comes up someone claims that they didn't get blisters. This is a dubious claim[10]. In 20 years, I've yet to meet a Candidate *at* SFAS who did not have blisters. This data point was a specific query during my 2-year long focused investigation. Blisters are universal.

You may ask yourself, if blisters are inevitable why invest effort into preventing them? The answer is well-aligned with rucking prep. Misery management. Your foot prep goal is to delay and minimize. You want to delay getting blisters for as long as possible during SFAS. Once you get blisters, your goal is to keep them from advancing to failure. What generally happens is that Candidates survive Gate Week relatively intact. But once Candidates transition to Land Nav week the movement and living conditions change. Your feet are likely to get wet, you will spend much more time on them, and the field conditions challenge proper hygiene. The blisters start here, and lazy Candidates don't treat them properly. The blisters grow or get infected. Before long Candidates are limping. This limping sufficiently alters your gait

---

[10] You often hear this claim from Candidates that didn't actually finish Selection. I'm not certain that your 10 days at Camp Mackall qualifies you to render much expert opinion on the matter. If you are the rare super-human who somehow avoided blisters, then you are the exception that proves the rule.

enough to cause issues in your joints, usually ankles, knees, and hips. It is not uncommon for a small blister on the heel to progress so quickly that within a day or two the once meager wound cascades into an infected foot or a painful hip joint because the pain of the blister forces Candidate to awkwardly limp to failure. Prep your feet correctly, manage your feet at SFAS, and learn to manage misery.

Getting your feet ready for the rigors of Camp Mackall takes time and you should approach the process deliberately. If you engage in this process correctly you can develop the habits to ready your feet for a lifetime of abuse, not just SFAS. Like all of my SFAS prep advice, I like to approach this with an eye towards simplicity. Simple is always better. I've seen lots of really bad takes about slathering on Preparation H, putting on nylons or dress socks, plus cornstarch, plus a thick pair of socks, etc. This is fine if you have a once-a-year roadmarch and you're just trying to get over the next event, but this is not a long-term strategy for success. I like to think in 3 terms: skin, socks, and boots. Let's start with your boots.

Boots

Some of the most common queries from hopeful Candidates are about boots. The SFAS packing list is quite clear. Boots must be in accordance with AR 670-1 [from the official packing list: BOOTS (all boots IAW AR 670-1, no GORE-TEX, no Temperate Weather boots, no buckles) as needed]. That's the guidance, that's it. This guidance hasn't changed in years, and I doubt that it will change during the lifetime of this book. The only specific characteristic that receives much scrutiny is a wraparound toe; that is a rubber toe cap. The Lowa Zephyr is a good representation of this characteristic. If the boot meets AR 670-1 guidelines, then you can wear it[11].

The good news is that there are so many great options for boots now and they are nearly mission ready right out of the box, it is really just a matter of user preference. Gone are the days of requiring *The Bootmaster* to apply his dark craft to your favorite jungle boots. Newer boots don't usually take a resole procedure well, so I

---

[11] Most Candidates report spending many hundreds of dollars in this process. This is not difficult given that good boots can easily cost close to $200 a pair. Socks are similarly expensive. So be prepared to spend a little bit, but you will find that it is money well spent.

recommend just finding the original sole you prefer and stick with it. In the past, boots required to be significantly broken in to be comfortable, so by the time you got the leather in the right condition the soles were usually too worn. Modern boots are ready or nearly ready right out of the box. Good boots can be expensive, but they are worth the investment.

There are a couple of criteria that you should keep in mind while looking for your go-to boot. First, you should do it in person. Online shopping is fine, but the wasted effort of delayed fitting and returns is unnecessarily complex. You should endeavor to do your fittings at the end of the day, so your feet are at their most swollen. Your feet swell, minutely, throughout the day even under normal conditions. Look for good ankle support, but not so much that it creates pressure on your Achilles tendon or restricts full flexion. You need the right foot support without binding, especially on the flex points. You need a roomy toe box, without letting your toes 'swim.' The sole type really doesn't matter but something with a heel (vice a flat sole) may help you on rope climbs. Too aggressive of a tread will clog easily, too little tread slips a lot. Remember that a soft sole will be comfortable for soft sand surfaces, but it may also transfer bumps and hard ground surfaces to your footbed. This is a minor inconvenience until you get a few hundred kilometers on your feet and every step you take is a business decision. Find what works best for you.

Two often overlooked boot components of proper fitment are inserts/insoles and lacing. You can change out your inserts and even get custom orthotics; there is no packing list restriction. You should consider having several pairs of well broken-in boots and rotate high quality, well-fit inserts to increase the longevity of the broken in boots without compromising support. I would hesitate to recommend any specific brand because like the boots themselves, there are so many excellent options, and the fitment depends on user preference. The SFAS packing list is very permissive – [*Insoles, Boots (Any Type)*]. Whatever brand you settle on, make certain you test them under the same conditions that you will experience at Camp Mackall. Figuring out that your super-gel insoles don't do well when submerged is something that you should figure out during prep, not while crossing Scuba Road.

Lastly, you would be amazed at the fitment differences that you can create by simply lacing your boots differently. The ability to lock your heel separately from the footbed or to relieve pressure from the top of the foot can be a simple few skipped lace holes away. The guide below shows a few of the most common techniques[12]. You would be well-served to know these prior to going to SFAS, even if you don't use them normally. Many Candidates find that the relentless abuse of SFAS requires an adaptive approach and the ability to relieve pressure on a bad blister or stress fractured metatarsal that you developed in-course can be the difference between select or withdrawal.

---

[12] Don't be that weirdo that uses 550 paracord to lace his boots. You're not in a survival scenario and you won't need to fashion a snare out of the 550 guts. Just get some normal laces and settle down, Rambo.

Standard Lacing | **General Swelling**

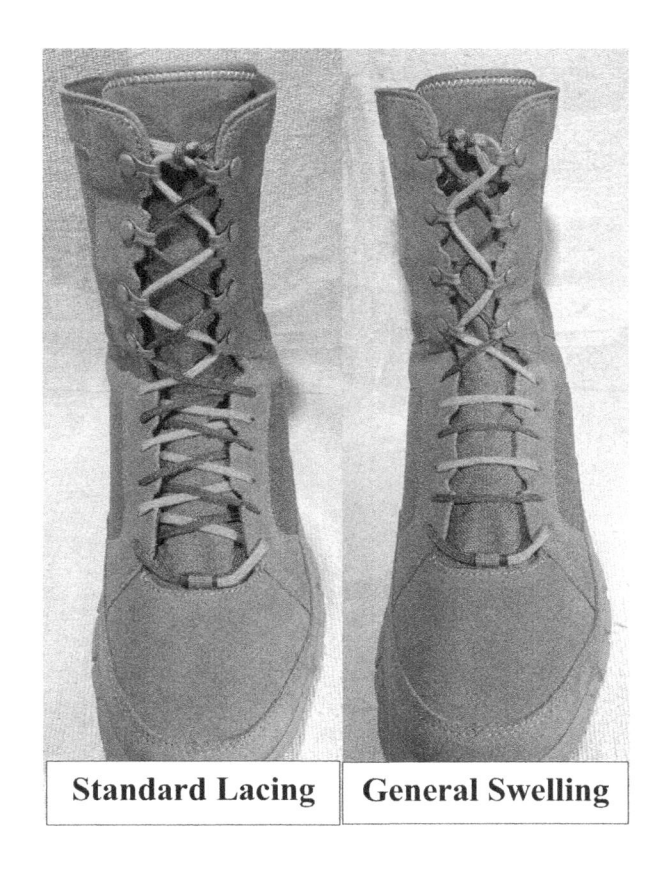

Mid Foot and Arch

Ankle/Heel Lock

Forefoot Pain | **Toe Nail Pain**

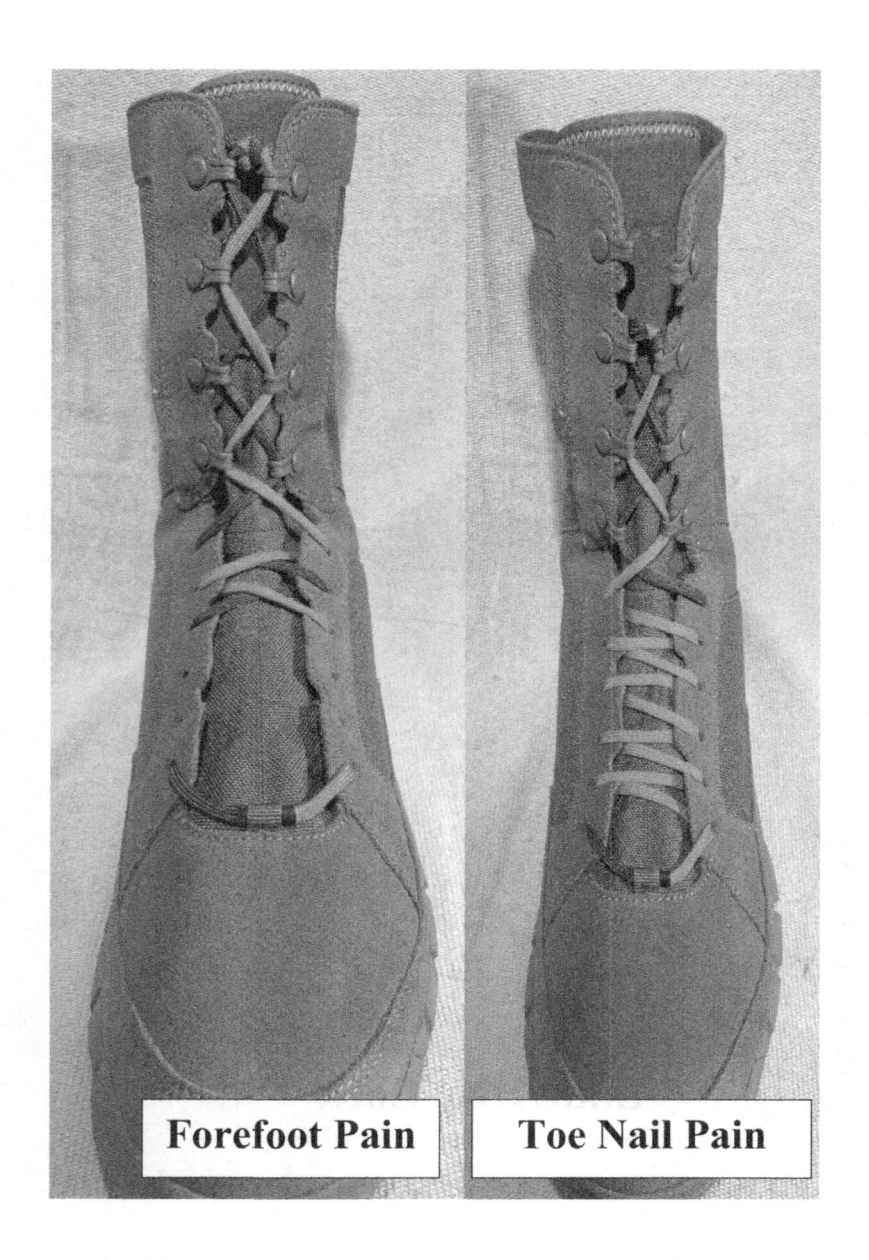

Let your boots dry completely between wearings whenever possible. You can speed this by removing the inserts and folding down the upper to allow better air flow. Avoid using direct heat. Rapidly drying your boots can lead to excessive shrinkage and can drastically alter fitment. Laying your boots in the sun or directly in front of a nice fire is tempting. The reality is that you are likely better off with slightly damp boots that aren't shrunken than you are with fully dried, but now altered boots. Add a small stiff bristle brush to your kit (on the SFAS Packing List [*Scubb Brush to clean TA-50 1 ea.*][13]. Leaving gritty sand embedded in the leather will hasten material failure and delay drying times. At SFAS, start your boot care routine as soon as possible after training has paused so that you can take advantage of natural drying as much as possible. Lazy Candidates want to sit around and tell war stories or eat MREs. Smart Candidates understand the priorities of work. Lazy Candidates don't last.

Socks

Guys seem to have oddly strong loyalty to certain brands of socks and are willing to defend their brand to the death. It's good to be passionate, but just remember that for

---

[13] Yes, I know that it says Scubb, that's what is on the Packing List.

every Darn Tough guy there is a Thorlo guy or a Fox River guy who is just as ready to do battle for his brand. Maybe the take-away is that there are lots of really good choices. It's all about finding the right sock that fits with your boot/insole choice and your foot type. If you sweat a lot, then you might need extra wicking properties. If you suffer from poor arch structure, then you might need some sort of compression support. If you are particularly sensitive or your preferred boot has a thinner sole, then you might need extra cushioning. There are lots of options for every requirement. The SFAS Packing List specifies [BOOT SOCKS, military colors only, no sock liners, no GORE-TEX, no waterproof socks, no toe socks]. Lots of room for personal preference.

Buy a couple of different pairs and try them all under different conditions to assess their worthiness. Aim for one solo pair, simple is better. I will never understand the mentality of a multi-sock system or having to put your sock on inside out because the seam irritates your foot too much. Just get better socks and stop beating yourself on a technicality that you've created. If you create a system where you need your inside-out 12% spandex liners and your custom hand-spun Tibetan yak wool heel-cushioned mega sox, then you're screwed when you don't have access to them. Don't just create a 'get through SFAS solution.' Create a 'get through life' solution that you can sustain while deployed or on mission. Then invest in yourself and get enough pairs of the 'winner' that you never have to decide between the best pair and another.

When you start rucking, build time into your workout to stop early in your movement and assess your issues. Note where you have hot spots. Adjust your laces appropriately, note what socks you have on if you need more or less cushioning, and note your load. Make adjustments and go again. Keep notes for reference later. What you should be doing is narrowing down all the variables so you can make informed decisions on which sock, boot, and insole combination is best for you. Don't guess, but don't overthink this either. Simple is always better.

Your Feet

Building your feet into a foundation that you can rely on to serve you under difficult conditions is not an overnight process. Some guys just have tough skin, and they adapt quickly. Some guys have never worn boots and will take months to build resiliency. Many will make months of progress just to abandon the principles of foot care and destroy their properly earned hardiness in a single poorly executed rucking session. The point is, start sooner rather than later. Avoid any bandages or moleskin when possible and opt for natural conditioning. The idea is to build calluses in the right spots at the right density.

Start with a simple assessment of your feet. Do you have an abnormal foot shape from wearing restrictive shoes? Years of wearing dress shoes can pinch your toes and unnaturally elongate your foot. If you primarily wear flip-flops you may have a wider natural toe spread that will feel crowded in standard boots. With this spread or pinching in the toe box you may experience blisters in odd places, like the tops of your toes. Look at your natural callusing. Curled toes can form raised calluses along the underside of your toes that will require additional attention. Angular heel bones or spurs can produce a pronounced bump in the back of the foot that will inevitably require attention. You might consider consulting with a skilled aesthetician who can give you a different perspective on your problem areas. They are normally focused on appearance not functionality, but the two are often linked and they can provide valuable insight. For bigger issues, consulting with a podiatrist is never a wrong move.

Foot powder, like socks, is another oddly tribal product. The once gold standard of foot powder was Gold Bond Triple Medicated. I can no longer endorse this product because they have updated their formula to eliminate talc and replace it with cornstarch. Despite being chemically treated, this old wives solution is literally feeding any potential fungus. Look for a non-cornstarch formula with zinc oxide and menthol. The zinc oxide is a skin protectant, and the menthol just feels good. Whichever product you choose, make certain that you are using it correctly[14]. Start with cleaning and drying your feet. Soap and water are best, but baby wipes are fine when pressed. For the powder, apply a light dusting to your feet and massage it in so that it gets in between your toes and full coverage of your foot. This ensures good coverage and confirms that you don't have any debris stuck to your foot that will cause issues once you don your socks. Don't dump it in your socks or boots as it

---

[14] Finding an American made non-corn starch foot powder is almost impossible. Our national litigious nature surrounding the use of talc has made it so. You can search out foreign made products, especially for the Asian market. I give pause recommending a specific product but look for one with zinc oxide to get started.

will just clump and once it gets moist it can create lumps that will prove troublesome. Less is more.

You should endeavor to develop simple solutions. If you rely on moleskin to keep your feet together, then what will you do when you run out of moleskin? Moleskin is a temporary solution for unforecasted issues. If you do use moleskin (it is authorized on the SFAS packing list), make every effort to apply it in a ring (or donut) around the perimeter of the hot spot to alleviate friction in the center. Avoid using a big piece to cover the entire spot. In training, if you need to apply moleskin to complete the workout, I recommend just ending the workout. Address the issue properly and save yourself more pain. A bad blister can delay future training events and your goal is to incrementally build resilience. In extremis, have a little baggie with some moleskin already cut to shape. Maybe include a small pair of sewing scissors to help trim to size. A sewing kit is a required item on the SFAS packing list[15]. You should include a blister kit as well. They are closely

---

[15] Sewing is a life skill and an SFAS skill. Learn how to do a few simple types of stiches so you can keep your uniforms and kit in good repair. Walking around with dangling roster number tapes is a sure way to send a clear message that you don't care. Selection is ongoing.

related. Each kit should have some quality, but compact, scissors. Try threading a needle with a thread cut by a knife and you will quickly become a scissor convert. The set that comes in most commercial kits is trash. Replace it. You could use that same set, or second set, in your blister kit to help trim your moleskin. The needles can be sterilized and used to drain blisters. With moleskin, less is more. Avoid using duct tape, another popular wives tale, to treat blisters. You can likely use it to spot reduce friction points, but now you have sticky adhesive tape that you must remove from a compromised piece of skin. You are likely to do more damage in the long run.

An unorthodox, but effective treatment can be to thread a needle with cotton thread. Sterilize both thread and needle with a little alcohol and poke that needle in, through, and out of a blister near the bottom of the bubble. Snip the needle from the thread and allow the thread to remain embedded in the blister. The exposed ends of the thread will wick the fluid from the blister and allow it to drain while not creating a large hole and further damaging the skin. This method takes time but can be extremely effective at draining the fluid while preserving the surface skin.

Avoid New Skin or similar products as they retard the creation of true calluses and delay adaptation. Remember our goal is to build the right thickness of callus in the right spot, at the right rate. If you want to encourage calluses you could try to wear your boots for a few hours a day without socks, but this might alter the true fitment of your boots so proceed with caution. You might also try applying a tincture of benzoin. Tincture of benzoin solution is a topical adhesive agent used to provide tackiness and enhance the adhesive property of tape. If you spread some on your feet and let it dry, over time this will help create tougher skin. If you have a bad blister, you can use it to 'glue' the blister shut but proceed with caution. This hurts like hell but can get you short term relief and that 'wound' now starts to create a callus. Too much callus can start to create new hot spots so regularly inspect your feet for ridges and overly thick pads.

Walking barefoot is not a recommended foot prep strategy. There is not one single event at SFAS that you do barefoot. Walking barefoot can actually increase your risk of blisters. You will likely build up calluses in locations that are not consistent with your boot calluses. Walking in sand can certainly help exfoliate your feet and build general resilience, but you also risk injuring your feet. A deep cut or abrasion can delay training for weeks. Leave this method on the shelf and concentrate on smart preps.

At SFAS, try to sleep with your feet elevated every night. It will help reduce swelling significantly and it seems to help keep your feet drier with the increased air circulation. I've seen more than a few Candidates that are so hobbled by swollen feet that they sleep with their boots on to avoid the painful swelling and subsequent struggle to don their boots in the morning. Don't do this. The normal routine is to wake up and pour your swollen feet into your boots. By the time you get to Team Week you will most certainly have blisters as well. It can take some willpower to get through the first ten minutes early in the morning while your feet get re-introduced to the day. Some Candidates call this process "lubricating," as in your blisters will start leaking enough to lubricate themselves into the correct position in the boot. That's an apt description. Jarring, but apt. A little more misery management.

Keep your nails well-trimmed. The prevailing wisdom and what you see from most professionals is to trim your fingernails slightly rounded but your toenails straight across. The idea is that this prevents ingrown nails. The problem is that this now creates a protrusion that rubs on your other toes and likely the boot. I don't think that this advice accounts for the foot swelling and extreme conditions that your feet are likely to endure. Rucking is different. My advice is to prep your nails by softening them (like after a shower) and trimming them with a slightly rounded shape. Don't create deep recesses by back-cutting them, just trim them slightly rounded. They will grow slower than your fingernails but inspect them regularly and look for any potential ingrown growth. Treat yourself to a pedicure occasionally and note how they manage your problem areas. Simple is always better.

When you are rucking regularly take some notes on how your feet feel and what socks, boots, and insoles you wore. Build a little database to help you isolate problems and develop long term solutions. The SFAS Packing List also allows for *Skin Care Lotion*. This is a generic moniker that allows some flexibility in exactly what type of lotion you choose. You might take advantage of this and select a product that is both lotion and a foot care product[16]. The key takeaways are to start prepping your feet early, be deliberate in your testing and evaluation of socks, boots, and insoles. Always seek a simple solution. Building good habits and taking care of yourself and your equipment/uniforms is critical to not only SFAS, but to good living once you get to a team.

---

[16] The packing list give you an opportunity here. You might consider a baby rash ointment or a lotion with some baby rash prevention and treatment ingredients. Most guys won't be too worried about a nice glow from well-moisturized skin, so some cocoa-butter lotion is unnecessary. But some 3-in-1 lotion that helps you treat swamp-ass will be helpful. You will see lots of bow-legged Candidates who are unaccustomed to the workload.

No Shit, There I Was...Boot Master

There seems to be a mystical culture around boots. This made sense in the olden days. One had to spend significant time and effort to prep boots properly. By the time you got the stiff black leather supple enough to ruck in comfortably, the soles were inevitably worn out. So, the Boot Master was born. You could take your boots to the Boot Master, and he would work his magic. He could remove the toe cap or the heel cup. He could eliminate the steel shank from your jungle boots. And he could give you a custom sole. The identifiable ripple sole sent a clear message that the wearer was serious about his rucking. Except they didn't always work.

I spent about six months methodically building up my feet in prep for SFAS, carefully rotating between several pairs of boots to break them in evenly and carefully monitoring my feet. I had the benefit of a prior service SF Physician Assistant in my Battalion Medical Station, and he trained one of the medics to be my personal foot valet. My daily visits were the perfect training opportunity. My feet were a finely tuned machine with faultless calluses and flawlessly trimmed nails. Impeccably groomed feet encased in immaculate boots. A simple foot care routine. A dusting of powder, one pair of issued socks, and jungle boot slippers. Perfection. Then I consulted the Boot Master.

About six weeks before SFAS, I took my best pair of carefully curated jungle boots to the Boot Master. I had a pair of those signature ripple soles installed to take my rucking game to the next level. They looked beautiful. They looked fast just sitting there. I couldn't wait to get them out on the trail. I laced up my new go-fasters and headed out for a quick 5-mile ruck. Just a little jaunt to kick the tires, so to speak. They felt awesome at first. By the end of the first mile they felt a little stiff. At mile two my feet were burning. I pulled a pit stop at the 2.5-mile turn-around point. My feet looked bad. Red, irritated, and chafed. I re-powdered and set back for home. By mile three I was questioning the boots. By mile four I was questioning my life. Despite my much-slowed pace and careful gait, I knew something was wrong.

When I limped back into the aid station and my *foot valet* peeled my sock off, his face told the tale of my error. I had completely torn off the bottom of my left foot. My carefully crafted calluses were gone. All that remained was raw, baby soft, and now bleeding and damaged skin. My right foot escaped mostly intact, but my left foot was thoroughly thrashed. And I had to report to Camp Mackall in just six weeks. We set about building that resiliency back up.

We started with getting rid of those boots. I don't know if they were badly done or if it was just a fluke, but those boots were never worn again. My foot was healed within a few days of careful bandaging, but they were too soft. We set about a deliberate regimen of tincture of benzoin applications. I referenced this method before, and I want to reinforce it now. I had amazing results and I've used it on others. It works. It is not a cure-all, and it won't work overnight, but it definitely works. Within a few short weeks I had built up my calluses again and never missed a day of training. At SFAS, my left foot remained almost entirely intact. Tincture of benzoin works.

There was also a group of weirdoes who showed up to SFAS and decided that this was the perfect opportunity to perform surgery on their boots. Idle hands make for the devil's work, I guess. In short order a whole crew of candidate-surgeons was ripping out toe cups and heel caps from their carefully prepped boots. I recall several of them complaining of new blisters and smashed toes just a few days later. SFAS is not the time to experiment with your boots. Boots now are so well made that there isn't much need for the Boot Master anymore.

    Prep smart, then execute. Simple is always better.

Life Lesson: Don't try to outsmart yourself. You don't have to reinvent the wheel for every task. Innovation is great, but not just for innovations sake.

Chapter 5 - How to Prepare for SFAS –

Land Nav Week

Land Navigation seems to be a waning skill in the military, and we are seeing the impacts of this in SFAS. The number of SFAS Land Nav drops is slowly increasing, with a rate of around 44% failure. The largest recruiting population for SFAS is the 18X population, so in many regards SWCS has outsourced the baseline navigation training to 11X OSUT at Fort Benning. The training during OSUT, even the expanded 22-week version, is lacking. A recruit is likely to only conduct a single solo traditional land navigation practical exercise before graduating OSUT. The general consensus is that OSUT graduates will receive more training once they arrive at their units. For 18Xs, their unit is the pipeline in SWCS, so they rely on the SFPC instruction. This appears to be sub-optimal given the SFAS success rates and reporting from 18Xs in the pipeline is that SFPC is more of a pre-selection rather than a preparation. For Candidates recruited from the operational force the challenge is obvious. Land navigation training at their units is of poor quality, if it even exists at all. Many units struggle to find a single instructor capable of providing competent instruction, especially in soft skill MOSs.

Officers tend to do much better at SFAS in general, and in land navigation in particular. This is likely because every commissioning source has a significant land navigation training and testing component as well as the general culture of fitness amongst junior Officers. When you don't know much about your job, as most junior Officers don't, you tend to focus on the stuff that you do know. Physical fitness is the obvious choice, but so is land navigation. The old trope about the lost Lieutenant is not well represented in the SFAS data as Officers do better than the general population.

Generally speaking, successful land navigation has three components: skills, confidence, and discipline. There is also a real value in repetitions. The more that you practice land navigation the more you refine your skills and build your confidence. Competence builds confidence. Ideally, Candidates will come well prepared, but the Cadre at SFAS provide excellent instruction and ample practical exercises to include Cadre-led terrain walks and area familiarization. There is significant emphasis on the SFAS rules to include time hacks, road crossing procedures, equipment requirements, and reporting processes. The high failure rates indicate that even this training scheme is insufficient. You would be well-served to supplement these efforts with your own preparation.

You should endeavor to participate in as much traditional military land navigation training as possible. Coordinate with adjacent units and seek to take advantage of their training, when tenable. You can also find civilian orienteering clubs in your local area that hold regular events. These events may not focus on the specific skills or equipment that you need at SFAS, but any time navigating is bound to produce transferable skills. Essentially, you need to understand three main fundamentals; how to calculate distance and direction, how to read terrain, and how to plan effective routes[17]. If you can develop multiple methods for calculating distance and direction, then you can reduce margins of error and move more confidently...and quickly.

My observations indicate that nearly every SFAS Candidate has the skills to be successful once they actually conduct the STAR. Many Cadre-led classes and multiple practice practical exercises over the exact terrain that the STAR is administered is excellent preparation. What they lack is the confidence and discipline. The physical and mental fatigue is real and essentially land navigation is

---

[17] Stay out of the draws.

rucking, but instead of moving on a trail or a road it is overland, and you have a known destination as a target. This should immediately reinforce the need for quality ruck training. This reinforces my statement that rucking performance is the most important predictor of success at SFAS.

So, what causes Candidates to fail SFAS land navigation? They lack situational discipline. They get fatigued and they abandon the principles of navigation. Know where you are, know where you are going, and know how you are getting there. Plan good routes that are efficient and avoid obstacles[18]. Increase your pace when the terrain and your routes allow it so that when you encounter unavoidable obstacles you have sufficient time to negotiate them correctly. I've yet to meet a Candidate who intended to cheat. What normally occurs is one of two things: They get a little disoriented and try to 'navigate harder' in the hope that they will figure out where they are, or they chose a bad route and get mired in a swamp and lose precious time.

If you train land navigation enough times, you will inevitably get lost at some point. Your mind will wander, and you'll simply lose track of where you are. It happens to

---

[18] Stay out of the draws!

even experienced navigators. When this happens, you need to stop moving. Return to your last known point and begin anew. Logically, if you begin a route at a known point then the task becomes to simply transfer that known navigable point along your intended route all the way to your final point. Move from terrain feature to terrain feature or intersection to intersection. What you use as your enroute check points is a function of your effective planning and what the terrain allows. Candidates get a little disoriented and don't return to their last known point, even if it is just a few hundred meters behind them. Instead, they drive forward under the false pretense that they will figure out where they are along the way. By the time they figure out their location, if ever, they have burned too much time. Many end up getting on a road to make up time and get caught, road kills. They didn't start the day intending to cheat, but they put themselves in that position by being undisciplined.

LAND NAVIGATION

PHOTO CREDIT K. KASSENS, USAJFKSWCS

Baseline Skills

Understand that nothing that I write on these pages is enough to prepare you for Selection. You need reps. You need practice and you need to get lost. You need to get lost so you can understand how to get un-lost. Learn to never *stay* lost. What does that mean? It means that you need three pieces of information to effectively land navigate. You need to know where you are, where you are going, and how to get there. At Selection you will be given the first two, where you are and where you are going. The start point and the end point. It is up to you to figure out how you are going to get there. That's called route selection.

For most novice land navigators, they take the start point and the end point and they mark them on the map. Then they draw a line between the two points and connect the dots. Then they lay a protractor over that line, and they calculate a distance and a direction. For many, that line becomes the route. If that line says the distance is 3500 meters at 170 degree azimuth, then by default the route is to walk 3500 meters at 170 degrees. If that line happens to cross over some swampy terrain or a draw, or it forces you to cross multiple roads, or it forces you through thickly wooded areas then so be it. That's the route. If you deviate from that route then you might get lost, right? Look at the straight-line route in the below graphic. Notice that it crosses multiple draws, swampy areas, and obliquely intersects multiple roads. A horrible option[19]. But that's not SFAS. SFAS is a start point, an end point, and *you* get to pick the route. You get to be in charge. You get to decide how you are going to connect the dots.

---

[19] STAY OUT OF THE DRAWS!

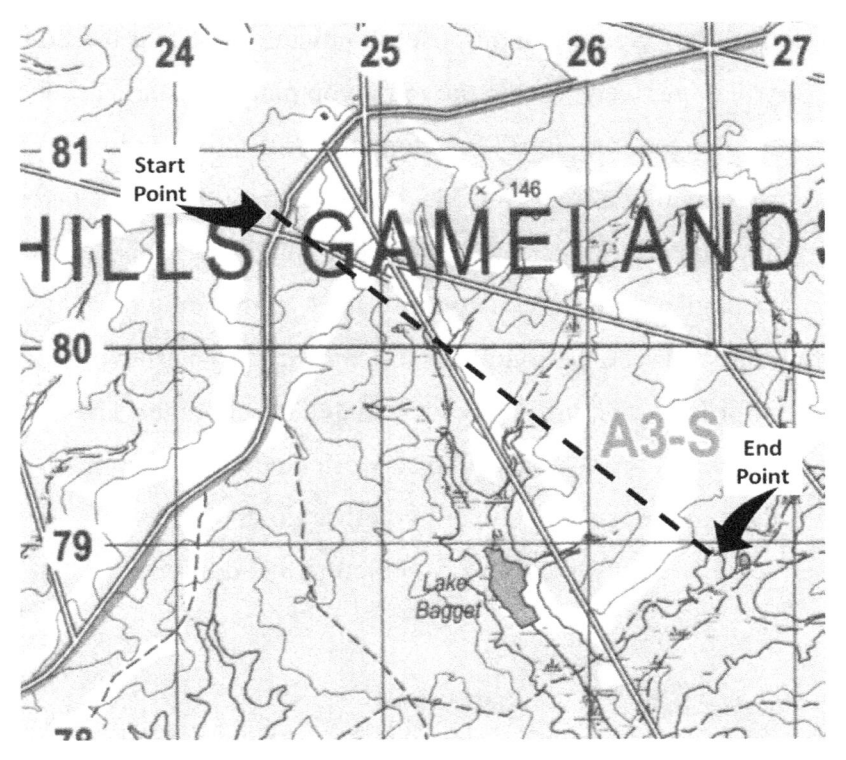

START POINT, END POINT, AND A STANDARD STRAIGHT-LINE ROUTE

Route Planning

Your task then becomes to pick the best route possible. Plan the route so that it takes you around the swamp. Select a path that avoids the draws. Instead of constantly crossing the roads, use them as navigational aids. You are in charge of yourself, so take charge. If you get to pick the routes, then why would you purposefully pick a route that forced you to get wet? Maybe the route you plan that skirts around the danger areas is 1000 meters longer. But on the longer route you don't get wet. Maybe the extra 1000 meters takes an extra 15 minutes to traverse, but you didn't walk through a draw and risk getting stuck in that draw for hours. Maybe the route was less direct, but your new route afforded you more easily identifiable checkpoints. These checkpoints allowed you to stay oriented and you didn't have to guess as much. Did you lose time, or did you actually gain some time? Subtraction by addition.

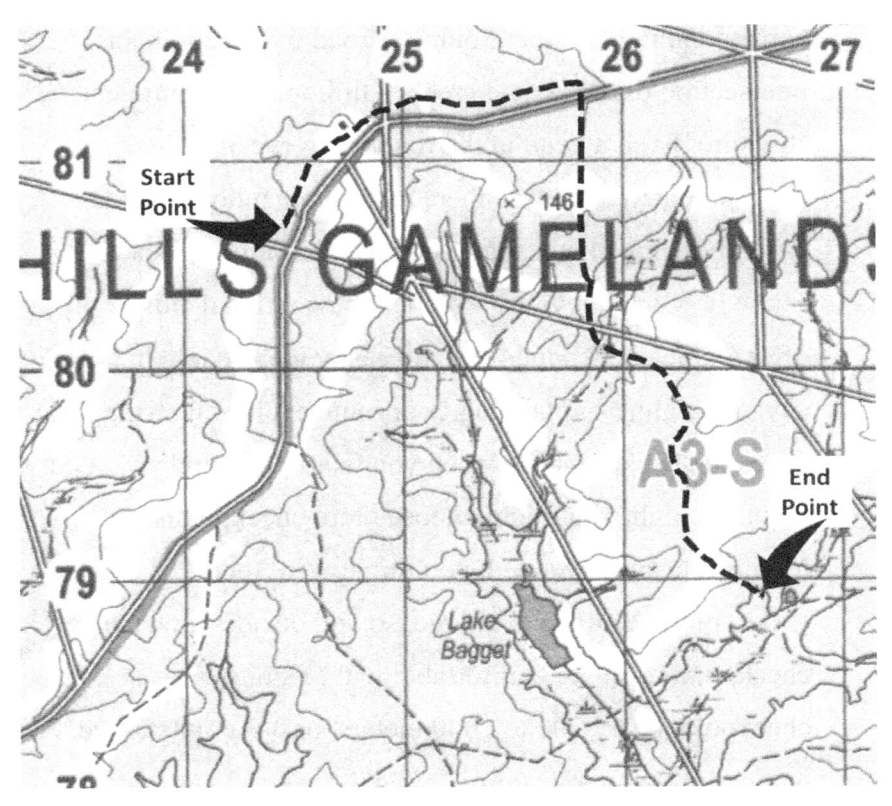

START POINT, END POINT, AND AN OBSTACLE AVOIDANCE ROUTE

Check Points

Let me introduce you to the idea of a checkpoint (CP). Some trainers will call them Rally Points (RP). Others will call them Rendezvous Points (RV). For this discussion we will use them interchangeably. All they are is a feature that is identifiable on both the map and correspondingly on the ground. A road intersection, an intersection of a road and creek, a hilltop, or a prominent structure. Even a bend in the road can serve as a checkpoint. At Selection there are prominent open areas, called bowling alleys, that dominate the training area. These bowling alleys are excellent navigational aids as they offer ample checkpoints that are easily identifiable in both day and night. So, planning your route entails choosing a series of checkpoints between your start point and your end point. At night you might choose more checkpoints. If the terrain is heavily wooded, you might want more checkpoints. You are in charge, so you decide how many checkpoints you are comfortable with. I suggest a checkpoint every 700 to 1000 meters, or 3-5 points based on our 3500-meter example.

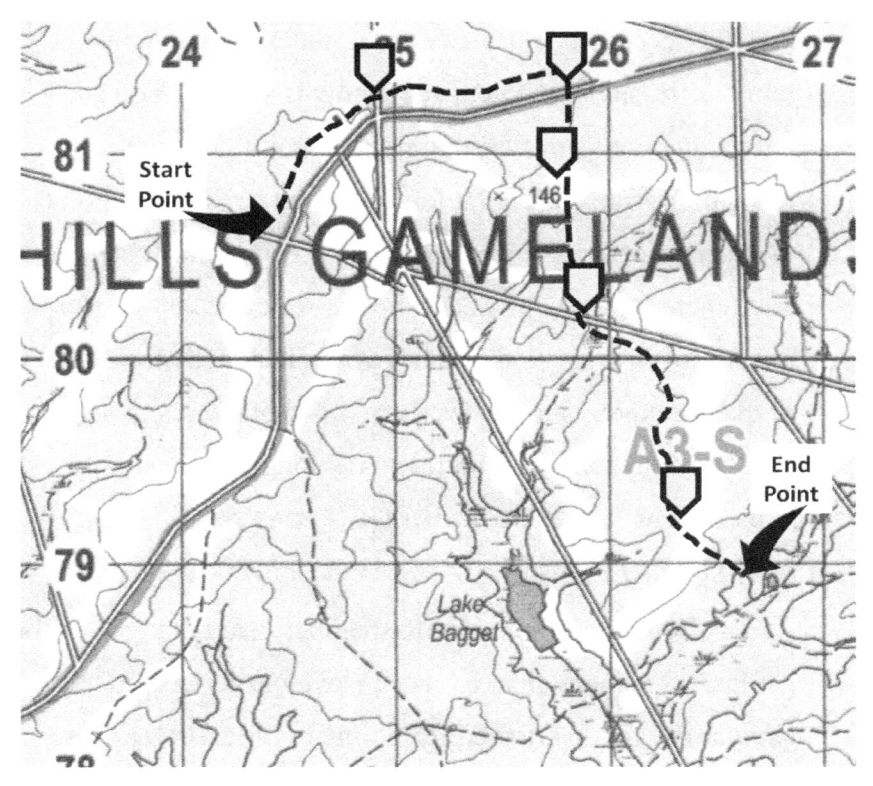

OBSTACLE AVOIDANCE ROUTE WITH CHECKPOINTS

    Your navigation task then is not to go from the start point to the end point, rather your task is to go from your start point to your next check point. Now instead of navigating 3500 meters, you are only navigating a mere 700 to 1000 meters. You just decreased your margin of error by nearly 80%. When you arrive at your first checkpoint and you confirm that it is the one you plotted, that now becomes your new start point. Your end point remains the same, but your route now is shorter. You are not navigating from start point to end point. You are navigating from start point to checkpoint, from checkpoint to checkpoint, and from checkpoint to end point. It is not a 3500 meter movement. It is simply a series of 700-to-1000-meter movements. And because you selected the location of the checkpoints that cleverly avoided obstacles, you don't have to contend with them any longer. You simply bypass them[20]. You are in charge.

Getting *Un-Lost*

Here is the cool part. Remember I said that you will get lost? It is inevitable. Everyone, even experienced navigators, get disoriented from time to time. If you chose those checkpoints wisely then the farthest that you have to go to get *un-lost* is 700 to 1000 meters. When you realize that you are lost and you can't get oriented, then just turn around and walk back to your last checkpoint. Don't wander aimlessly in the false belief that if you just "navigate harder" your location will magically reveal itself. Be disciplined, turn around, go back to your last check point, and start that segment of your route again. If you

---

[20] STAY OUT OF THE FUCKING DRAWS!

"navigate harder" you could be doing so for hours. If you are disciplined enough to simply return to your last checkpoint it might take 20 or 30 minutes. Subtraction by addition.

Handrails

Navigating effectively is knowing where you are (start point), knowing where you are going (end point), and knowing how you are getting there (route selection). You are given the first two and you are in charge of the third. You can plan that route effectively by plotting good checkpoints. Your challenge then becomes staying oriented along the 700 to 1000 meters. Do you recall the roads, creeks, and bowling alleys that helped us identify good checkpoints where they crossed paths? Because they are linear, they can serve as boundaries for our movements as well. If you plan your route correctly you can plot that route so that it has one of these linear terrain features on either side of your movement corridor. You are not allowed to be on the roads, but they are going in the direction that they are going, and you can use that to your advantage. If you are moving to the south and a road along your route is oriented north-south, then you can be in the woodline and look at the road and 'follow it," without actually being on the road itself. That is called "handrailing." Much like you would use a traditional handrail to guide you down a flight of steps, you can you use the road to handrail along your route. You are not on the road, but you can still use it for navigation. Likewise, your end point may have a linear terrain feature 'behind' it, that is the linear feature intersects your route. In this case, that linear feature becomes a 'back stop' to your movement.

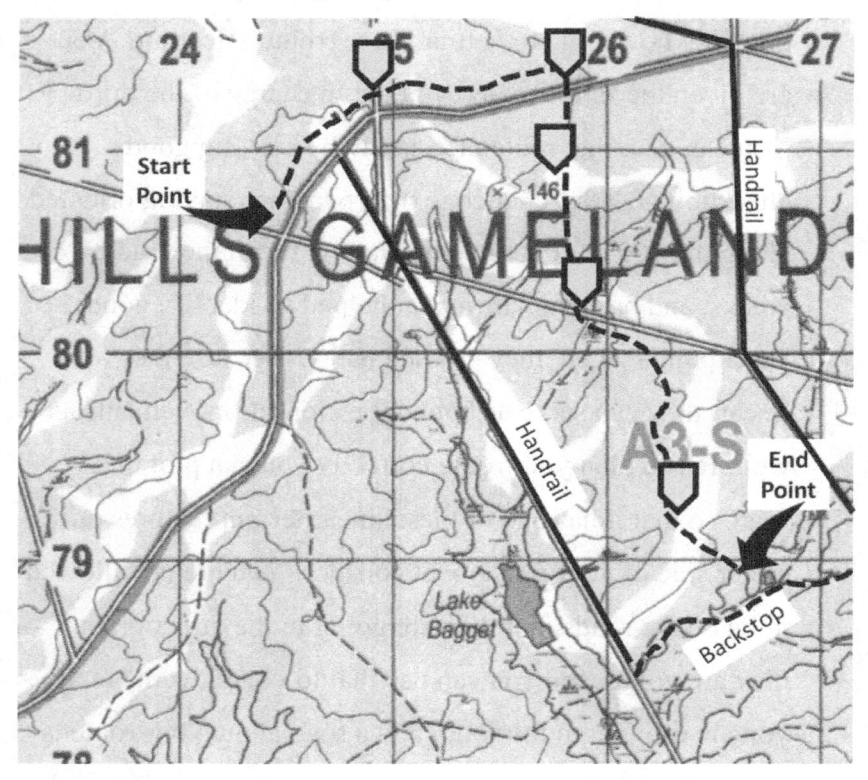

A Complete Route with Checkpoints, Handrails, and a Backstop

Good navigation then isn't just start point, end point, and distance and direction between the two. It becomes an exercise in checkpoints, handrails, and backstops. Staying oriented along that route. Managing your time effectively. Following all the rules. At Selection, you must do all that with a ruck on your back. Day after day. Night after night. Mile after mile after mile. And there is more. We haven't even covered search procedures, RV procedures, or cadre interactions. It becomes an exercise in managing your cognitive load.

EVERY IMPORTANT DECISION THAT YOU MAKE AT SFAS,

YOU MAKE WITH A RUCK ON YOUR BACK.

PHOTO CREDIT K. KASSENS, USAJFKSWCS

Managing Cognitive Load

If you can't think straight, then you can't act straight. At SFAS you are being judged on both, simultaneously. You need to develop systems and processes that allow you to structure critical repetitive tasks so you can stop actively thinking about them and free up that cognitive capacity for real time processing. Because land nav is concurrent cognitive and physical load, it is a prime opportunity to use this technique. Let's use *RV procedures* as an example.

We previously discussed RVs as interchangeable with CPs and RPs and this heuristic is still valid. You should execute this procedure at regular intervals along your route, so CPs, RVs, and RVs are all compatible. Essentially, you should develop a set process that you do every time you come upon an RV, CP, or RPs. If you execute this procedure regularly then you eliminate a significant cognitive load. Every SFAS class, many students lose maps, scorecards, end even weapons. Almost every student loses their orientation at some point. By establishing simple

RV procedures, and being disciplined about executing them, you can turn this liability into an asset. Here is how it works.

Develop a systemic process in which you inspect/account for all of your sensitive items, key equipment, and cue yourself for action. Have you ever noticed how you perform a pat-down inspection every time you leave your house? You pat yourself down to account for your wallet, keys, glasses, and of course your phone. You don't even think about it, you just do it. You inherently recognize that these items are so critical, that you developed your own RV procedures. This is why you rarely misplace your phone or your keys. This habit is so engrained that pickpockets use it to identify which pockets your valuable items are in when they assess potential victims. You just instinctively show them by patting yourself down.

If you develop a similar RV procedures system, you eliminate the risk of misplacing key equipment. For example, I follow the *Head, Shoulders, Knees and Toes* model. Every time I come to an RV, CP, or RP I pat myself down from my head (do I have my compass and maps on a dummy-cord around my neck?), shoulders (do I have my ruck and load bearing equipment?), knees (do I have my weapon?...I know it's not related to my knees, but you get the point), to my toes (I physically look on the ground to both identify my terrain feature and to ensure that I didn't drop any of my equipment). I also use this as an opportunity to clear my mind and reset my thinking.

When I follow my RV procedures, I am certain that I never lose my equipment. If, for some reason, I have lost some equipment I can limit my search between where I am and the last position that I conducted my RV procedures. If I planned my routes correctly and I put a waypoint every 700 to 1000 meters, then I can focus my search in that specific area, instead of the entire route that I have walked. I conduct these RV procedures before I leave my start point, at every RV, CP, or RP, at every prominent navigational aid that I encounter (even if it wasn't pre-planned), and just prior to arriving at my end point.

My RV procedures are just an intuitive part of my navigating and because I don't have to worry about my gear, I can use that cognitive capacity for actively navigating. When I am closing in on my end point and I cue myself, I am mentally preparing to engage in a very critical information exchange with a Cadre member or a point sitter. I am preparing myself to be assessed. I've made certain that I have all of my equipment and I have sorted myself out to make certain that I am presenting the most competent version of myself possible. When I have that exchange, I can respond with a clear mind, and I can take decisive action. I gain this huge advantage by simply establishing a habitual RV procedure. I am managing my cognitive load.

You should apply this thought process to many of the functions you will routinely engage in at SFAS. Every time I put my ruck on, I grab it by the frame and lift it up over my head and slide into the shoulder straps. I do this not just because it looks cool, but because it helps me check to make certain that my ruck is packed correctly. If I was digging in my ruck and forgot to secure it correctly or have some loose straps, this procedure will reveal these deficiencies to me. This system served me throughout SFAS. When I am nearing the end of my land nav route and I am searching for my point, I always look to my left first. I am right hand dominant and therefore I naturally drift to the right while walking. Logically, my point should be on my left. I don't have to think about it, I just do it. If you develop enough of these heuristics, then you can free up significant cognitive capacity. A massive advantage.

Your land navigation prep then is simple. Learn the baseline skills; how to plot a point, how to measure distance and direction, and how to plan routes. Get as much practice as you can so that you can build competence and confidence. Stay disciplined and follow all of the rules. This definitely is not easy, but it certainly is simple. There are some modest tools that you can acquire to help you along the way. A good map case is essential. Ziploc bags simply don't last and if you want to risk your chances of selection on a disposable sandwich bag then that is on you. I recommend the Seal Line large or medium cases. I rarely endorse specific products, but this one earns the testimonial. These are the gold standard and I have several that have stood the test of time. You should get an issued lensatic compass. It is heavier, bulkier, and uglier than a quality commercial equivalent, but it is absolutely bombproof. They are accurate, easy to read in day or night, and incredibly reliable. This is the compass that you will use at SFAS so you might as well build your skillset with what you will be tested on. You should also invest in a good watch, not a GPS, or an altimeter, or a digital compass. A watch. Everything that you do at SFAS is timed and all of your land navigation is a solo event. Measuring exact time is critical to your success.

Have you noticed that I keep focusing on simple, logical, and practical preps? No complex systems. No subscriptions and proprietary formulas. No secrets and you don't need to know what the exact events or the exact standards are. There are no shortcuts. Faster than the strongest lifter, stronger than the fastest runner. Slow is smooth and smooth is fast. If you develop a reliance on overly complex conditions to perform well then you are building yourself into a hot-house pansy, only able to thrive under perfect conditions. Special Forces is an austere capability. SFAS is downright Spartan. Humans are more important than hardware.[21]

---

[21] I should note that I run an SFAS Land Nav prep program. I started this endeavor in direct response to the critical deficiencies in DoD provided land nav training and the rising land nav failure rates. Learn more at TFVooDoo.com.

Chapter 6 - How to Prepare for SFAS – Team Week

HOW DO YOU PREP FOR THIS?

PHOTO CREDIT K. KASSENS, USAJFKSWCS

On the micro level, one of the best preps for Team Week is taking care of yourself during Land Nav Week. Wear your gloves, especially when navigating draws. Little cuts and abrasions can fester and if you have damaged hands for Team Week, you will not fare well. An ounce of prevention is the recipe for success. When you are completed with your daily training you need to avoid the tendency to just plop down and commiserate with the other tired, sore, and grumbling Candidates. This starts with good hygiene. Keep yourself clean so that you can keep yourself healthy. Focus on your feet. Forty-four percent of Candidates who seek medical attention do so during land navigation and the most common complaint is blisters and abrasions. Take care of your feet, you're going to need them. Take care of your boots. They are likely wet, so get them dried properly. Flash drying them next to a fire is inviting disaster. Burnt boots are common. Maintain your uniforms and gear. Make necessary repairs and clean everything as needed. Eat as much as you can and drink plenty of water. You just completed a rigorous physical task, essentially an hours-long cross-country speed ruck. You should be conducting recovery as such. Likewise, you should conduct pre-workout mobility drills, when tenable, every day. You want to go into Team Week as fresh as possible.

On the macro level you want to focus on getting strong. The same way that we have a good understanding of what it takes to build rucking performance (field-based progressive load carriage, usually 2-3 times a week focused on short intense sessions) we also know how to build strength. But what kind of strength? Strongman powerlifter strength or triathlon strength? Explosive compound lifts or calisthenics and body weight exercises? The simple, but unsatisfying answer is yes. Yes, to everything. Stronger than the fastest runner and faster than the strongest lifter. This is what makes Team Week prep so challenging. How do you prepare to move thousands of pounds, quickly, across varied terrain, under different conditions for near endless hours? High carry on your back and shoulders. Low carry with ever diminishing grip. Pulling ropes and pushing jeeps. Relatively stable dead weight and the shifting mass of water cans or pails. All of the rules matter.

The answer is that you train everything. Lots of modalities. Everything is a priority, so nothing is the priority. You need a program that allows for adaptability but follows the progressive overload principle. We need specificity but we just established that we need to train everything. A conundrum, to be sure. There is no available data for just how strong you need to be, so absent of any evidence lets apply some logic to this problem set. We can still maintain our simplicity mandate, but we need to be more mindful now. More on mindfulness later.

In my faster/stronger paradigm we do have some benchmarks. We know that you must be able to run 2 miles in at least 15:12 for the PFA, so 7:36 a mile. We also know that SWCS recommends a sub-40 minute 5-mile run. That is a minimum benchmark. We also know that run performance is a strong indicator of Selection success.

Similarly, we know that there are some specific rucking time benchmarks (Teplitzky, 1990)[22]. They might seem a little dated, but my experience confirms these times to be excellent benchmarks for performance. It's fair to say that faster is always better, both rucking and running. We can use these various data points to establish a logical speed benchmark: 6–7-minute mile run targets and 12-13 minute mile with 55-pound ruck targets. Your strength training regimen should then focus on getting you as strong as possible without sacrificing speed to the established targets. You might think those run and ruck times are quite fast. They are. SFAS is a high barrier to entry and being a Green Beret is to commit to the lifestyle of a professional tactical athlete. PC or Beret.

We now have our speed benchmarks, and we know that we need to be strong, so let's work on some strength benchmarks. Your strength goals should focus on building lean body mass as this is a key indicator of selection success (Farina, et al., Anthropometrics and Body Composition Predict Physical Performance and Selection to Attend Special Forces Training in United States Army

---

[22] Read this study and pay attention to the hypothetical ruckmarch cut-off analysis. This is where we can extrapolate some time benchmarks. In short, faster is better. But this study can show you likelihood of selection as it correlates to specific times.

Soldiers, 2022)[23]. 1 would assess that you need to be stronger than you think. No successful Candidate has ever said that he wished he wasn't so strong. But there is some nuance that we should tease out here. We know that a prerequisite strength level is required before you will likely see significant rucking performance gains, so our first strength benchmark is pre-prep (Scott, Shaul, & McCue, Ruck Deep Dive: Study #2 - Ruck Training Adaptation, 2015). You should be able to lift your body weight in the bench-press and 1.5x your bodyweight on the squat. Once you reach this pre-prep level you can start to work on your actual rucking[24].

Your strength prep should include some calisthenics because the specificity principle mandates that we train for the Hand-Release Push-up and Pull-ups. But calisthenics alone are not sufficient to build the sort of strength that you will need for Team Week. You can monitor your calisthenics requirement by programming a diagnostic PFA into your regimen every few weeks, as discussed earlier. Don't get caught being really strong but not having the

---

[23] Bigger guys do better, but not fat guys. You need lean muscle mass, and you need to be able to move that mass quickly.
[24] You can start rucking right away, but you are more likely to get injured and you won't see the returns on your investment until you build this prerequisite baseline strength.

correct technique to pass the PFA. There is a little gamesmanship to these events and if you don't train for them, you will just be the strongest PFA failure in the non-select tents. Based on your diagnostic results you can adjust your training regimen as needed. If your scores are low, then add more calisthenics. If your HRPU and PU numbers are good, then you can focus more on traditional weightlifting.

Similarly, your strength prep should include the sort of specificity that mirrors Team Week events. I won't describe all the events, but The Sandman can serve as a valuable model. Carry this incredibly heavy and awkward weight on your back and push this heavy dilapidated jeep as fast as you can for miles and miles. It is not just the straightforward task though. It is also the positioning and the rest breaks. Now instead of just carrying this heavy apparatus on your back, it also involves picking up this awkward apparatus off the ground and up over your head so that your teammates can shimmy their rucks underneath it. Hundreds of times a day.

So, compound lifts seem to fit this model. Barbell Back Squats, Barbell Deadlifts, Barbell Bench-press, Overhead Shoulder Press, and Bent Over Barbell Row should be the foundations of your strength program. Any exercises that work your hand grip strength should be heavily emphasized as well. Many critics will claim that the bench-press is a poor prep for SFAS because after the PFA you don't do any chest press specific assessments. But the bench-press is a staple of general strength building and is an indicator of overall upper body strength. The specific motion of anchoring your feet into the ground, leaning in, and pressing the jeep up that hill is eerily like a bench-press. Albeit, you are in a compromised and awkward position, but it is a chest press, nonetheless. I recommend that it remain in your SFAS prep. In general terms, pick up heavy stuff from the ground and put it up over your head. Repeatedly. High weight and high volume. Progressive overload. Simple. You will note that I have not recommended or critiqued any specific program. Like socks and foot powder, guys seem to be oddly territorial about their prep program preference. I'm convinced that much of this tribalism is the sunk cost fallacy. "Program X is the best program because I chose X and I'm no idiot!" I would instead look for simple plans that follow the foundational principles (specificity, progressive overload, and adaptation) and SFAS specific adaptability that I have described[25]. Find or build a system that works for you, addresses all of the specificity and strength requirements, builds ruck and run capacity, and allows you to be faster than the strongest lifter and stronger than the fastest runner. Simple is always better.

Knots

The topic of knots seems to be of particular concern for many Candidates. The Army likes to make knots some sort of exclusive task in many schools. At SFAS there is no 'rope corral' or times test, but good knots, and a good Knot Guy, can be immensely helpful. If you can't tie a knot, then tie a lot. The Mongolian Booger Knot will work, but you might struggle to untie it. I think there are 5 knots that you need to be proficient in for SFAS. The square knot, the clove hitch, the bowline, the water knot, and the sheet bend.

The Square Knot

---

[25] Again, I'm not throwing shade on any particular program. There are some that are pretty good, but they seem overly complex…and expensive. The free SWCS and THOR3 prep programs do not meet the criteria I have outlined. The inside story of these program's development is interesting, but the results fall flat. It is best to follow the principles that I have listed, but don't get too cute. No need to build your own Sandman.

The square knot is ideal for tying two pieces of rope together to create one long rope. The ropes that you will be issued in SFAS are complete or nearly complete, so overall length is rarely an issue. What seems to happen is that they are so long that they become cumbersome. You end up with odd length running ends and they drag underfoot or become entangled in the apparatus. The square knot will allow you to dress up those loose ends and it is easy to untie. I've seen teams miss their hit times because they couldn't *disassemble* an apparatus after movement was complete.

THE SQUARE KNOT

The Clove Hitch

The clove hitch is ideal to anchor a rope or webbing, especially around a pipe. The pipes are the foundation of many apparatuses and the axle of every apparatus that rolls. Learning how to properly anchor a line can set that line for the duration. A poorly anchored line will fail and cause inevitable delays. Essentially, it's three wraps of rope around an object, with one of the wraps creating a lot of friction for holding the rope in place.

THE CLOVE HITCH

The Water Knot

The water knot is an effective way of securing two lengths of webbing together, creating one long length.

Basically, it's two intertwining overhand knots. When dressed correctly, this knot distributes friction across all the surfaces, offering a secure hold.

WATER KNOT

The Bowline

    The bowline is a way to tie a fixed loop in the end of a rope (end of line bowline) or in the middle of a rope (inline bowline) that cannot loosen, shrink, or expand. A bowline knot creates that loop, and it can function as a hand or foothold.

The Bowline

The Sheet Bend

The sheet bend is used when tying two ropes of varying diameters together or a rope and webbing. One material cinches down on itself, over top of the other material, creating a strong connection.

THE SHEET BEND

A General Timeline for Prep

Time	Domain	Measure of Performance
1-2 months	Establish baseline strength	Benchpress .75x BW, Squat 1x BW, Deadlift 1.5x BW, 75# Grip R/L, Trapezius Build, HRPU, Pull-ups
	Establish baseline aerobic capacity	90 minutes continuous at 75% max HR
	Establish baseline skills	Rope climb, balance, sled drag, tire flip
	Establish baseline mobility	Full ROM on all joints, hips to ankles on squat, no shoulder impingement
	Establish baseline rucking technique	Understand pace, gait, and shuffle
1-2 months	Build functional strength	Benchpress 1x BW, Squat 1.5x BW, Deadlift 2x BW, 100# Grip R/L, Trapezius Build, Introduce compound lifts, maintain PFA
	Build baseline aerobic capacity	90 minutes continuous at 75% max HR
	Build baseline mobility	Full ROM on all joints, hips strength thru full ROM, bilateral shoulder ROM (extension/internal rotation and adduction, and flexion/external rotation and abduction)
	Build baseline skills	Rope climb and mantle, balance and shift, sled drag, tire flip
2-4 months	Build rucking fitness	Build pace, weight and distance to 12 min mile, 55#, and 5 miles 2-3 x wk, 1 x 12 mi/mo. No new blisters
	Build functional strength	Build BP, SQ, & DL. Grip to 150. Max compound lifts, Maintain PFA
	Build aerobic capacity and speed	90 minutes continuous at 75% max HR, intro speed work
	Maintain mobility	Full ROM on all joints, hips strength thru full ROM, asymmetric ROM
	Maintain skills	Rope climb and mantle, balance and shift, sled drag, tire flip

One of the most frequently asked questions is how long it takes to adequately prep for SFAS. You should be able to surmise from the last 90 or so pages that there is a massive amount of knowledge, skills, attributes, and fitness that you need to address. If you are out of shape, then you need to account for this. If you are in good shape but have never rucked, then you need to account for this. If you have young kids at home that require your parenting, then you need to account for this. There is no single answer to this question, but I would offer the following as a potential model. This is more performance based than time based.

In the first 1-2 months you need to establish your baselines across strength, aerobic capacity, skills, and mobility. There is no need to start your prep program at a full sprint because you want to super-charge your performance. In fact, starting too fast and too strong will hinder your progress. For example, you will note that there is no rucking in this phase. We know from the literature that you need to build a pre-requisite strength level before you will see any significant rucking performance gains. So, this phase of your prep establishes that strength baseline. Just work bench-press, squat, and deadlift. Work on your grip strength, work on your PFA, and build your traps so that you can properly support a rucksack in the next phase.

We are also building aerobic capacity. Do as much running milage as you feel comfortable with but stay in Zone 2 and work to go up to 90 minutes per session. Nice easy runs to acclimate the body to the workload. Work on some skill stuff too. Learn the different ways to climb a rope. Work on your balance. Learn to drag a sled and learn to flip big tires. You don't have to do these skills at any real intensity, just learn the skills. Finally, establish a rock-solid mobility program. Call it yoga or Pilates or movement prep or stretching, it doesn't matter. The goal is to get flexible and extend your range of motion as much as safely possible. This phase is meant to assess and prep for the greater and more intense workloads of the following phases. You need to start to understand how your body responds to specific stimuli and what domains you need additional coaching or focus for. Slow is smooth, smooth is fast. Train smart. Simple. Not easy, simple.

The next phase of 1-2 months you should build on the foundation that you built in the first 2 months. Most importantly, you will start rucking now that you have established the baseline strength requirements to get the proper returns on your investment. You should work on your pace so you can start to instinctually understand what a 15 versus a 14 versus a 13-minute mile feels like. Keep your weight low in the 25-35 pound range. You can work on your gait, and this is where you will start to understand the artistry of a good shuffle. Don't rush to failure. Slow is smooth, smooth is fast. For strength, keep working on your bench-press, squat, and deadlift numbers as indicated. Get your grip numbers up as well. Every Candidate cites grip strength as a critical measure of performance. You should also introduce the compound lifts now. Again, you don't have to be an Olympic level lifter yet, just learn the movements correctly.

You should maintain your running Zone 2 aerobic capacity, but now you need to balance out your total milage with your rucking endeavors. You should be balancing your recovery, your workload, your nutrition, your rest, and of course your performance. You are a data scientist now and your experiment is you. Maintain your rope climbing, balance, and ancillary skills as in the first phase. Think about the Nasty Nick and the CRA and use that as your specificity goal. Continue your mobility work and pay particular attention to your shoulder health. Remember the Team Week stuff where you are picking up this heavy apparatus and lifting it up over your head and onto a ruck. Those awkward, off-balance, compromised base motions. Keep your shoulders limber and strong.

In the next phase we can start to put this all together at pace. If you prepped smartly then you have 2-4 months of setting the conditions for success. You should feel comfortable in the gym. You understand all the movements and you can do them safely, at pace. You should now start pushing some significant weight. Stronger than the fastest runner. Stronger than you think you need to be. Strong on every axis and strong with a compromised base. You should be able to really ruck well now. You spent the last few months building technique, understanding pace, and building those trap muscles to support some abuse. You can shuffle. You are a fledgling misery manager, and you are skilled at that quick little shrug movement that allows you to settle your straps into the proper position to extend the session. And your feet are getting hard. You are strong enough that you have the musculature to build speed and add weight.

Continue with your Zone 2 run training but manage your overall mileage in concert with your rucking. You should start to incorporate speed work which can help you reduce mileage but still train specificity. You should know exactly which boots and socks are your go-to pieces and picking up a ruck should just be a thing you do, not a thing that you dread. Maintain your rope climbing, balance, and ancillary skills and start to put these sessions together with a ruck or other event so you can start to understand how this physical exhaustion impacts your cognitive functioning.

You should really manage your mobility heavily now and be very mindful of injury. If you timed your prep properly then you should peak as you report to SFAS, but you can't go injured. You're not that tough, and nobody at SFAS will care if you used to be really fit. You need to perform and if you're injured you just won't be able to. Track your rest, hydration and nutrition carefully. This general guideline is 4-8 months in duration. If you started at a high level of fitness and prep then you might be able to accelerate it a little. If you had to balance a rich family life, manage an injury, or work around a difficult work schedule then it might extend longer. Everyone responds to stimuli differently so keep good training logs and be smart about your individual programming.

Prep Conclusion

That is a comprehensive, but sufficiently generalized description of SFAS and how to prepare. These recommendations are based primarily on the evidence as documented in the vast body of peer-reviewed literature, other credible media sources, and my specific experience and observations. Be smart, be fit, and be a good person. Know how to navigate both the physical terrain and the social environment. Learn to lead and be led. Lift heavy things, tie good knots, and move quickly over varied terrain for time. Be a renaissance man. I would recommend starting your prep as early as possible but learn to manage the intensity so that you don't burn out or injure yourself. If you have to balance a family life, then this prep might take you years. Do not go to SFAS if you are injured. You will not have time to recover and you will not make it. Try to time your prep culmination just prior to reporting to Camp Mackall. That's it. Simple. Everything is a priority, so nothing is *the* priority. This is why SFAS is such a high barrier to entry. Sixty-four percent of Candidates will not be selected. Know what you are getting yourself into. PC or Beret.

SFAS PREP SUMMARY CHEAT SHEET		
	Event	**Prep**
Gate Week	PFA	Monthly Diagnostic PFA
		General strength program
	Cognitive Assessments	Be well-read
		Communicate effectively
		Be well socialized
	Timed Runs	Run Prep
	Timed Rucks	Ruck Prep
		Foot Prep
	Nasty Nick	General strength program
		O Course PEs
	CRA	General strength program
Land Nav Week	STAR Exam	Ruck Prep
		Foot Prep
		General strength program
		Land Nav classes
		Land Nav PEs
Team Week	Team Events	General fitness program
		Communicate effectively
		Be well socialized
		Ruck Prep
		Foot Prep
		General strength program
	LRM	Ruck Prep

No Shit, There I Was…The Silver Bullet

A common cause for concern for many Candidates is the packing list. It is a rightful concern because if you don't have all of the required items listed, then you could get dropped. There is also cause for concern that if you bring the wrong things the cadre may take notice. That Grey Man notion again. Many candidates go the opposite direction and bring bizarre items. Bedsheets and a pillow are authorized, but you will see purple fuzzy crushed velvet sheets or an Elmo pillow. Star Wars sheets are more common than you would expect. A book is authorized as well. I will never understand the motivation that drives an SFAS candidate to bring a book about SEALs to Camp Mackall. There are more than enough books about Green Berets, just bring one of those.

One of the most frequent questions is about compression clothes. The packing list specifically prohibits any compression shorts or shirts. I am often asked why that is and the medical staff at Camp Mackall told me that compression clothes, shorts in particular, can mask serious injury and might place candidates at increased risk. I've seen this firsthand. When I went to Selection, we had a

Candidate who tore his hamstring. At that time compression shorts were authorized as part of the PT uniform, so many Candidates had them. This Candidate was able to mask his injury and continue training because of his compression shorts. He kept training to the point that he tore his hamstring entirely, and he ended up with a grotesque deep black and purple bruise that covered his entire leg from his ass to his knee. It was nasty. But wait, there is more.

We know what this grotesque injury looks like because this Candidate ended up getting "trauma naked." During Team Week we were enduring one of the endless movements under load and we were closing back in on the cantonment area nearing the end of the day. There was a particular sharp bend in the road that we quickly learned marked the last final push towards rest. Teams would naturally sense the nearing completion and pick up the pace a little. My Team had just started our final kick and as we burst around the bend in the road, we were greeted by the sight of that black and blue leg injury. We could see it because the medic had taken the candidate, gotten him 'trauma naked', and placed him ass-up draped over his ruck sack. And that rucksack was firmly planted smack dab in the middle of the road.

I'm no medic, but I'm certain that I could have adequately examined his leg with a less revealing examination position. But 18Ds are known to be a little off sometimes. Eager to perform minor surgery for almost anything and possessing of the sort of bedside manner than would earn few stars on Yelp. But this medic took it the extra mile. He decided, in order to complete a comprehensive exam of his patient, that he needed a temperature reading. But not just any temperature reading, a rectal temperature reading...The Silver Bullet. So, as we rounded the corner we were greeted by a clearly injured candidate, bare-assed and in the most compromised position one could hope for, with a Silver Bullet thermometer sticking out of his ass. The medic added to the scene by snuggling up close, leaning on his patient, and sort of draping his arm across him in a nice comfy pose.

We of course had to stop. Our progress was blocked by the now immobile patient. The medic gave a quick assessment of the injury and we dutifully nodded along trying not to notice the ass thermometer. He noted that the compression shorts were about the only thing holding his leg together. Our Cadre asked him if he needed to call in for an evacuation vehicle or something else and the medic nonchalantly replied, "Oh no, not at all. I was done ten minutes ago. I was just waiting on you guys for a little show."

Never trust a medic with an agenda. And don't wear compression shorts.

Life Lesson: You will get lots of nicks and dings during SFAS prep and most definitely during Selection. Learn the difference between being hurt and being injured. Just like the compression short ban at SFAS is designed to protect against further injury, sometimes you can be your own worst enemy by training through injury or not pushing yourself through being just hurt.

Chapter 7: Can't Tell Me Nothin', Can't Show Me Nothin'

I wrote this book as a baseline description of SFAS and a broadly applicable baseline prep plan. It is most applicable for active-duty guys as your experiences are already well-shaped and commonly understood. The other two categories in follow-on chapters will address the younger guys and the older guys. First, let's cover some frequently addressed issues from the active-duty guys.

Active-Duty Guys

I'll say active duty, but I really mean in-service. Guard, reserve, sister-service, prior service, whatever. This group seems to be the hardest to manage. You guys know just enough to think that you know it all, but most of what you know is hearsay. You seem to have retained counsel through a licensed Barracks Lawyer, and your barrister failed Selection[26]. And you are just proud

---

[26] Many of you are Barracks Lawyers yourselves. You should decline your own counsel. Some of the worst advice that I have seen comes from failed Candidates or guys on active duty who

enough to not take real constructive criticism to heart. Can't tell me nothin', can't show me nothin'. Your selection rates are likely an indicator of this condition; active-duty enlisted soldiers are selected at the lowest rates. Read slowly and take notes.

Let's start with the big stuff: the CGs Unwaiverables. The SF Regiment has learned, and relearned, that there are some things that simply aren't compatible with Special Forces. As I write this book, we seem to be re-re-learning them. Moral waivers. In a world of endless exceptions to policy there are three things that should bar you from service in the Regiment. You shouldn't be allowed to have any felony, any alcohol related incident, or any domestic violence history. In 2010, SWCS was managing a large influx of serious disciplinary action across the Regiment to include tab revocations. For those unaware of how rare tab revocations are, it used to be perhaps one a year. Prior to 2010, I can only personally recall one, ever. Circa 2001, a Master Sergeant in 7th Group was busted for assaulting his 14-year-old stepdaughter. The charges were sexual assault, but he

---

have years of bad habits. SFAS is a uniquely demanding environment and the techniques that you use to get through that once a quarter 12 mile/35-pound roadmarch are not transferable.

raped her. He is a rapist. We revoked his tab. Publicly. He was previously by all accounts a 'good guy' and well liked. But we officially removed his tab and we PNGed him[27]. He was convicted and hauled off.

But that was it. The only one that I can recall. I'm certain that there were others, but my point is that this stuff was rare. Big Boy rules because we acted like big boys. We specifically screened for these sorts of abnormal behaviors, or even indicators of abnormal behavior. The SFAS psychological evaluation is pretty good at this. But we also used to have some other behaviors that we looked for. There have been various versions of the Unwaiverables over the years. At one point it was just DUI conviction, other years it included any alcohol related incident, but there were markers, indicators of deviance.

During the GWOT surge, around 2005-2007 timeframe, MG Parker, then the Commanding General of SWCS, lifted those restrictions. Anyone in SWCS at the

---

[27] Persona Non Grata is a Latin phrase that means "an unwelcome person." It is a legal term used in diplomacy to indicate that a diplomat is not acceptable or welcome in a country. We use it as a verb to indicate that we have disowned someone.

time can tell you that these were desperate times. Cadre were told that they would double the output but that standards would not be lowered. Anyone with even a scintilla of intelligence knows that this isn't possible. Cadre were stretched to the absolute limits. It damn near broke the system. I've seen the data and the overall selection rates weren't that different than historical averages. But we were sending more Candidates overall, so total output did go up slightly.

So how did they recruit more Candidates? They entirely lifted the 'moral' standards: no alcohol incidents, no domestic violence incidents, and no felonies. This doesn't mean that every guy who has a DUI is a child rapist, not by a long shot. But the cumulative effect of lifting this barrier was predictable. So circa the 2010 timeframe we had a whole slew of tab revocations and serious disciplinary actions across the Regiment. Something like 20 over a ~2-year period. Major General Sacolick, the SWCS Commanding General, had us research the issue, and ~90% were guys that had one of these issues in their records, but got waivers during recruiting. There was a clear correlation.

So Major General Sacolick established the CGs Unwaiverables. No alcohol related incidents, no domestic violence, and no felony. No waivers. No exceptions. If it's in your record, then you did it. The circumstances aren't important. But as new leadership takes control and political landscapes shift, so does our tolerance for deviance. If there is any lowering of standards, it is here in the recruiting process, not in SFAS. If you happen to be one of the unlucky souls that requires a moral waiver, then be prepared to work hard to submit the packet through your recruiter. But also, be prepared to be denied and be prepared for someone at SWCS or USASOC to come to their senses and reestablish the unwaiverables part.

Beyond the Unwaiverables, your prior MOS, rank, branch, assignments, or deployments don't matter. Yes, you are a product of those things, but the Cadre don't know or particularly care about them. There is one standard at SFAS and only your performance matters. I've seen experienced veterans from the Ranger Regiment quit while zero-deployment, non-badged logisticians get selected. Don't make the excuse that you aren't X or you don't have your Y keep you from dropping a packet. Likewise, don't think that because you are Z that you will get a pass[28]. The only thing that matters is your performance.

Review the earlier chapters of performance expectations and preparation guidelines carefully. But I would offer you this benchmark to meet before you go to selection. 1) Be injury free; you will not have the luxury of recovery between events. Unlike your real-world position, you don't get to pick and choose when you can take a break or skip an event. 2) Be able to do 75 hand-release pushups, 10 dead-hang pullups, and a 12:30 two-mile run. Be able to do 12-13 minute mile rucks with 55 pounds pretty much indefinitely. 3) Be able to bench press your body weight, squat 1.5 x your bodyweight, and deadlift 2 x your body weight, at a minimum. Be stronger than the fastest runner and faster than the strongest lifter[29].

Family Life

A word about wives. This topic comes up often and SOF has a reputation for being a marriage killer. I think that this is a self-fulfilling prophecy. Enter this world with your eyes wide-open, but there is nothing inherently more

---

[28] Most guys self-select before they ever even drop a packet. Stop making excuses and let us decide. You will get a fair shot and you never get to yes by staying silent.
[29] Yeah, I know that I keep saying this. Maybe there is a reason why.

dangerous or divorce inducing about being a Green Beret. There aren't many secret missions that you can't talk about. There is a strong argument to be made that this career is safer in that Green Berets get the very best training, the latest and greatest equipment, and we are surrounded by the most competent warriors ever produced. We do dangerous things, and we go to dangerous places, but we are masters of chaos, and we never gamble. Risk yes, but we never gamble.

Becoming a Green Beret should be a family decision. You don't necessarily need your wife's permission, but if you make this journey against her will, she will become resentful. Especially when you start your SFAS prep. You need a willing partner. There are some incredible perks for being a Green Beret. You have better home station assignment stability and predictability. You don't have to deal with junior Soldiers and the inherent issues of junior Soldier's wives. You get special incentive pay and allowances. Your deployments are usually well coordinated, and they are generally shorter than your conventional counterparts. But earning your beret and living this life cannot become an obsession.

You need to account for your family in your prep. Be home for dinner and weekend birthday parties. Don't slack on your parenting duties. Train as much as you can on your time and protect your family time jealously. Make some training a family event. Throw on a ruck and take the dog with you. Push a stroller and bring your wife. Make it a Team Week event and have her slowly antagonize you with a running time countdown. If you have a wife and kids, it may take you longer to train up. But it also gives you a reason to train even harder.

Don't ever forget why you are undertaking this endeavor. You want to lead a purposeful life. A life worth living. Eventually, you will retire. We all do. If you make your entire life about earning and keeping your Green Beret, then you will end up a lonely man. No matter what patches you earn, what medals you are awarded, and what deployments you tallied it ultimately all ends. The day after you retire the Army will forget all about your accolades and you become just a number. The Brotherhood lives on with you, but the Army quickly forgets. Your family, the reason why you did this from the start, is still there[30]. The Army

---

[30] I have had an amazing career, but my greatest accomplishment is that we are still married, and we raised two great kids.

forgets, the family endures. Make certain that you never forget that.

But being Active Duty should give you a significant advantage. You are conditioned to the hurry up and wait culture. You are familiar with the non-sensical, and Camp Mackall is an upside-down world. You can task organize and you can get stuff done. Cadre won't care, but other Candidates will certainly know who you are, and they will look to you for your experience and situational awareness. If you are smart you will take advantage of your status. You also have access to gyms and physical therapists and nutritionists. You have access to land nav training and training areas. You have every advantage that you could need.

Chapter 8: The Lame, the Blind, the Mute, and the Crippled

Older Guys

The maximum age for SFAS is 34 years, but this is one of the most often waived requirements. Older active-duty guys tend to have a more refined understanding of the limitations of age, so most of the interest from older guys comes from civilians. Males likely hit their peak fitness in their mid-30s, but SFAS is unique in that there is almost no time for recovery. You don't control your rest, your nutrition, and your load management. Older guys just don't adapt well, so don't wait too long.

You older guys seem to often focus on two key aspects, telling people what you do in your civilian life and asking if you can still do those things while serving. Not surprisingly, many of you are looking to serve in the National Guard. You have established a certain status and lifestyle and you don't want to abandon that life entirely. Your civilian accomplishments don't really mean anything. Your degrees, your licenses, your titles, and your collegiate athletic endeavors don't mean much. They are an important part of your past and might give you valuable experience, but in and of themselves they don't mean much. It seems weird that many of you seem to include these accolades in your dialogue. It won't make a difference with recruitment (even if the recruiter tells you it does), peers will think its awkward if they keep getting reminded of them, and the cadre certainly won't care. Only performance matters.

Because of your experiences many of you come into the process with a little baggage, or hardware. There seems to be an endless option of ETPs (Exceptions to Policy) for all sorts of medical hardware, conditions, and limitations. About the only thing that I've never seen get a waiver is a heart condition or a TBI. Everything else is a matter of 'it depends.' It depends on your specific condition, how diligent your recruiter is, and the exigencies of your medical screener. Your best bet is to just find a recruiter that you can trust and give him what he needs to put your packet together completely. This is true for young guys trying to navigate the ADD medication limitations to older guys trying to figure out how many screws they need to have removed. I recommend that you read Army Regulation 40-501 *Standards of Medical Fitness* for some good specifics. But you only get to "yes" if you put in your packet. Nobody can really predict how it will go so there really isn't much value in worrying about it. Submit your packet. Shoot your shot.

National Guard or Active-Duty

This is a common topic amongst the *older guys* population so I'll include it here, but it's an interesting study in the different operational culture of the National Guard. Individually, National Guard SF guys are virtually interchangeable with active-duty guys. At the unit level, National Guard units are not as good. It is entirely unreasonable to expect a unit that has significant time restrictions to be as synchronized as a unit without those restrictions. I recognize that that's a loaded statement and it's not a judgment of value, it's just a jumping off point. I'll frame this discussion in two parts, at the individual level and at the organizational level.

Individually, an active-duty SF guy goes through the exact same pipeline as a National Guard SF guy. It's the exact same SFAS, the exact same Q course, and the exact same Advanced Skills courses. There is no National Guard standard at SFAS (something I've heard multiple times that is flat out false), and you're not more likely to 'squeak by' because the Guard is "hurtin' for bodies." One standard. But National Guard does boast a slightly higher selection rate. This is likely due to the pre-screening process. SFRE does a good job of weeding out the truly unprepared and many Guard units do a good job of mentoring guys through the process. It varies widely from unit to unit and state to state, so if you have really specific questions, it's always best to reach out to that specific unit. Guard SF units are actually really responsive to queries, and you can easily find their contact info online.

The same higher SFAS pass rate isn't mirrored in the Q; there is no statistical difference between active-duty and National Guard. The same goes for Advanced Skills, no statistical difference. Slots for Advanced Skills (and other schools) are allocated to every unit and managed at the unit level, so the opportunity to attend courses is the same. But it's a challenge to balance the additional time away from your civilian life to dedicate to an outside endeavor. It is just a logical tyranny of time. Many guys report that it's easier to get school slots in the Guard because many of the guys simply aren't available, no competition for slots. So, if you're flexible and willing, in the Guard you can sometimes go from school to school to school. But I've never seen active-duty guys struggle to get schools either. The only school that makes its max quota consistently is MFF.

I can't cite any vetted data, but it is not uncommon for National Guard SF units to have a good amount of prior active-duty guys. It obviously varies tremendously from unit to unit, but it's a popular off-ramp for active-duty guys looking to pursue other goals but that still want to 'touch the magic.' It's tough to dedicate so much energy towards an endeavor and just walk away cold turkey. National Guard guys are just as patriotic, just as motivated, and just as dedicated as active-duty guys. But they have a lot of stuff like an entire civilian career, all of the friction of aligning disparate training opportunities, and limited resources that compete for their readiness. Some can overcome this, but many can't. The Army isn't going to pay for your local gun range fees, or your skydiving lifts, or your gym membership. The results are not unpredictable.

There is no 'most common' or 'preferred' civilian career for National Guard SF guys. Law enforcement, defense contractor, and first responder are natural fits, but the civilian career field is as diverse as you can imagine.

Executive, business owner, contractor, doctor/nurse, IT, whatever. There is no restriction, but you do have to keep in mind that serving in the Guard (not just SF) can be a burden on your civilian career. It's illegal to discriminate based on service, but some employers are logically reticent to invest in an employee that they could lose for a good chunk of the year. It's also not uncommon for guys to travel for hundreds of miles to drill with their unit. I know guys who live on the East Coast but drill on the West Coast and vice-versa. The cost of that travel is rarely reimbursed. Some units consolidate several drill periods, so they only meet a few times a year, but for longer periods. I have very frequently had Guard guys say that they work way more than they get paid for (a common theme for leaders in all Guard units, but prevalent at all ranks/positions for SF) so it very much becomes a thing that you do for the sense of duty rather than any real financial gain. This is very unit dependent and there are no guarantees so just understand the potential burden going in.

There is the enduring urban legend of the 'Guard Bum' that floats from deployment to school to deployment and essentially stays on active-duty as an individual augmentee or 'gun for hire' so to speak. I've known a couple of guys that did just that for a decade or more. But those are a rarity and becoming rarer as we reach a more manageable steady state operational tempo post-GWOT. It's more likely for low density MOSs like 18D and 18E because you can't deploy an ODA without a medic or a communicator. 18Bs and 18Cs are more prevalent so they are in lower demand for individual augmentation. So being a 'Guard Bum' is possible, but very unlikely.

At the individual level there is virtually no difference. At an organizational level it's a different story. First, you need to understand how different the National Guard can be from both active-duty and the Reserves. It obviously varies from state to state, but Guard units are notorious for being undermanned, under-resourced, and chock full of nepotism. The 'good ol' boy' culture is alive and well, and that's not always a bad thing. But in a community as purposefully insular as SOF, overlaying an additional stratum of insulation can be burdensome. I've heard a myriad of tales of favoritism and partiality that you just don't get on active-duty. It happens on active-duty, but it can be on a whole different level in the Guard. This is especially true for Officers and the competition for commands and key billets. Very political and unpredictable.

At the unit level, like the individual level, the tyranny of time is real. Active-duty units struggle to maintain the full depth and breadth of skillsets that are expected of them. Individual training, collective training, unit level stuff. There is just so much that needs to get done. This is doubly so for MFF and dive teams. Now imagine trying to do that on 80% less time, where your training resources and personnel are geographically dispersed. It's absolutely unreasonable to expect that they would be the same. Officially, National Guard SF units are assigned the same mission sets, the same core competencies, the same deployment types, and the same operational expectations. There is no official taxonomy of preparedness, but there is certainly an understanding of this dynamic. This isn't to say that National Guard units are seen as 'less than,' or deficient, or Junior Varsity, but there is absolutely a recognition of the likely limitations.

In the early days of the GWOT a National Guard SF unit was tasked to serve as the CJSOTF HQ in Afghanistan. Not the entire unit, just a reinforced staff to run the CJSOTF HQ mission for a standard duration. This was a couple of years into the fight so there was plenty of time to get up to speed. During the pre-deployment certification process it was clear that they were not up to the task. They were undermanned, unorganized, and overwhelmed. There were certainly some guys that were individually well-prepared, but organizationally they were simply not up to the task. Most guys were great dudes; willing but just not able to pull it all together. We had to delay the RIP/TOA and rally a robust augmentation team of active-duty guys in key billets just to get them out the door. We have since invested a tremendous amount of energy and resources to avoid repeating that scenario (and have largely accomplished our goal) but that institutional memory runs deep. That's a reality.

So officially National Guard and active-duty are the same. This is certainly true at the individual level. If you were to put 10 National Guard and 10 active-duty guys into a room together, you likely wouldn't be able to tell them apart. But if you put 10 National Guard units and 10 active-duty units, even at the ODA level, into a training scenario/full mission profile then you would almost certainly be able to notice differences. Sometimes only subtle differences, and sometimes only discernable to the trained eye, but you would almost certainly see the differences. I would be remiss if I didn't note that there are some missions that are actually better suited for National Guard teams. If the mission is to train a partner nation police force and you have an ODA made up of majority LEO, then they would be ideal for that mission. If you had an infrastructure mission and you had an ODA with a bunch of engineers, contractors, and craftsmen, then they would be ideal. But that's a product of personnel, not a component of the organizational dynamics.

So, now you know. I should note that I never served in the Guard so I'm not an expert and I likely missed some key details. It can also be very unit/state dependent so there likely isn't one single correct answer. I should also note that many of you may be putting the cart waaaaay before the horse. You might dedicate a few months to prep and see how you respond physically to the rigor the mission set demands before you hang your whole future on the prospect.

No Shit, There I Was...The Long Tab

This isn't a referendum on National Guard guys, it just happens to be about a National Guard guy. During the first part of the Q course, we were enjoying a nice leisurely breakfast in the Camp Mackall dining facility. Enjoying a leisurely meal is a rare event, and it was about to become even more rare. We had a fellow student, a National Guard guy, in my class who had earned the *President's Hundred* tab. The *President's Hundred* is a badge awarded by the Civilian Marksmanship Program to the 100 top-scoring military and civilians in the President's Pistol and President's Rifle Matches. It is a big, long tab that is higher in precedence than the Special Forces long tab. Its size and shape are an anomaly. It is exceedingly rare and an honor to be certain, but also a liability for some.

During our meal a Cadre noticed the Soldier's tab and asked him about it. It was a totally normal and congenial query. For some bizarre reason this student decided that that was the day that he was going to test himself. He loudly replied, "Well, that's a long tab that you'll never get!" You could hear a mouse fart. Everyone froze and suddenly became very interested in the tray directly in front of them. The Cadre, safely sequestered in their little Cadre section of tables stopped mid-bite and just stared. You could cut the tension with a knife. The enquiring cadre just smiled and calmy replied, "Well, alright then," sat down, and enjoyed his meal. But you could sense a shift in the dark side of the force.

We quickly finished our meals and made for the exit. Lots of whispered, "Holy Shit, did you see that?!" and "Who was that guy?!" We were in shock. Nobody would even acknowledge the 'long-tabber.' You just knew that the Cadre's calm retort wasn't the end of it. A great injustice had been done. Of course, we didn't enjoy an indoor meal for the remainder of that phase of instruction. Most meals were replaced by MREs, and the rare hot chow was eaten outside. A couple of weeks later we finished our training and as we were standing in formation to board the trucks back to Fort Bragg and continue the Q course, we looked around at our peers. We were missing one particular person. We were a 'long tab' short.

I don't know what happened to that guy, but it always pays to understand your operational environment.

Life Lesson: Respect is free; disrespect can cost you everything. Know when to shut the fuck up. It's a fine line between being funny and being a smart ass. Don't blur that line or you'll pay for it. We love a guy who can crack a joke, we hate the guy that takes it too far.

Chapter 9: The Young and the Restless

If you're still in high school, you should concentrate on being in high school. You can't go to SFAS until you're 21 for a reason. Get good grades, make lots of friends, play sports, and have fun. We make good Green Berets from good Soldiers, we make good Soldiers from good citizens, and we make good citizens from good people. You must be a good person first. So just work on being a good person.

Get good grades, not because you need good grades to get selected, but because we want smart people who can read and write. It shows a level of discipline, and it demonstrates a willingness to follow the rules. The modern American education system is designed to create moderately intelligent people who can become reliable workers into our industrial base. You should augment that by reading. Read a broad spectrum of topics. Check my notes below on the topic of Reading Lists. Read the classics and read modern novels. Read the bevy of great books that came out of the GWOT. But read normal stuff, too. Be intellectually curious. Learn to ask good questions. Be a critical thinker. Learn a little bit about a lot. Learn DIY skills. Learn to cook. Learn a language. Learn to sew. Learn to clean. Life skills.

Make lots of friends. There is no requirement to be an extrovert or socialite, but ~85% of those selected are extroverts (Bartone, Roland, Picano, & Williams, 2008). You need to learn to communicate effectively, both receive and transmit. Learn to give and take instructions. Learn to criticize and learn to take criticism. Learn how to carry a conversation and learn how to be quiet and really listen. Learn conflict management and how to be managed when you're being a problem. Have real life friends. Playing video games is fine but go touch grass more often. Learn to build a dangerous ramp and skin your knees. Build resiliency.

Play sports, especially team sports. This will help you build a good foundation of athleticism. It will also help you learn to take instruction, often critical and direct instruction. Coaches want performance and can have a unique way of telling you. Learn to be coached. Playing sports will help you build social skills, too. It doesn't really matter what sports you choose, but you might consider a contact sport like football because you should learn how to take a hit and be bullied. And you should learn how to deliver a hit and bully people. You might consider a combat sport like wrestling or jiu-jitsu. You aren't likely to score a melee kill in real life although it has happened, but you will get fit in a unique way, callous your mind and body, and cauliflower ears send an inimitable warning that you might be dangerous. You are years away from your pushups, pullups, run, or ruck times making any real difference. So don't worry about them. Be stronger than the fastest runner and faster than the strongest lifter.

Have fun. Be a kid. You have your whole life ahead of you and deciding to be a Green Beret is a huge commitment. It is also far more likely that you will fail than you will succeed. That's just the reality. Don't become one of those statistics that I cited earlier. Don't be fat, don't be dumb, and don't be unlawful. Historically, 64% of those who attend SFAS will not make it. If you have your whole identity and future built around this entire endeavor, then you run the very real risk of an imminent emotional crash. So, get your license, do kid things, and make mistakes. Don't catch a felony but understanding risks and learning to navigate tenuous situations is a good skill. Good judgement comes from experience; experience often comes from bad judgment.

Make mistakes but stay off drugs and severely limit your drinking. I am not making a moral argument here, but there certainly is a moral component to this. Stay off these substances because your brain is still developing. When you use these substances, you are altered. Sometimes you are altered temporarily, sometimes you are altered permanently. But you will be altered while your brain is developing. Because you will be altered while your brain is developing you run the very real risk of permanently impairing your proper development. And drugs and alcohol impair your judgment sufficiently that you are more likely to make an enduring legal error. The same can be said for prescription drugs. This is not medical advice, but there is ample evidence for the over-prescription of ADD, ADHD, and other SSRI medications and while many patients report remarkable results, these drugs stay on your medical records and complicate enlisting or commissioning. Don't refuse medical treatment for legitimate diagnoses but seek multiple options. Dial in your diet and exercise, manage your screen time, and learn to journal effectively before taking medications. Train yourself in good habits that help with focus and organization. Research ways to overcome ADD/ADHD to see if they work for you.

When you get closer to joining you should reach out to a recruiter. Don't put too much stock in your uncle's friend who knows a guy, or your buddy who plays lots of Call of Duty, or a random guy on Reddit. Understand that recruiters have a job; to recruit. So, they are going to sell you the best possible story and minimize all the negatives. They have quotas to fill and bosses to answer to. If it sounds too good to be true, then it probably is. If you aren't getting good answers, you can always talk to another recruiter.

Talk to your parents, too. If they have done their jobs well, they will be rightfully concerned. You have told them that you are interested in this thing. This dangerous thing. It is their job as a parent to protect their kids. But not just protect, also to prepare. If they did their jobs well, then you should know what the right path looks like and how to avoid the wrong path. Then tell them that it is time to let you walk that path. At a certain point you need to live your own life and you don't need your parents' permission. There are dragons to be slain so they should not be afraid to raise knights in a time of dragons.

The choice between National Guard or active-duty is entirely up to you as is the choice between 18X or a longer path towards being an officer. Nobody can make those decisions for you. It all depends on your individual goals, circumstances, and desired end state. This is veering pretty far from my original SFAS guide and prep mandate, but it is worthy of discussion.

A lot of younger guys don't want to wait until they are 21 to serve. They naturally look towards getting into SOF via the 75th Ranger Regiment. This is excellent prep for the rigors of SFAS, but you should know that the SF Regiment and the Ranger Regiment are entirely different worlds. The operational culture that I described earlier in this book is descriptive enough, but I could summarize this by saying a few more words.

The Ranger Regiment eats their young. A fledgling Ranger private has virtually no autonomy. You are not a Big Boy, and you will be treated as such. Harsh discipline, directive leadership styles, and a mission first mindset creates the world's premier light infantry unit. But it also does not have the luxury of accommodating your preferences. I can think of few better ways to build fitness, shape resilience, and acquire skills that will help you to get through SFAS. But you better be prepared for the specific rigors of that lifestyle. Eat or be eaten.

18X

The 18X program is designed to get guys right off the street into Group; Special Forces is historically undermanned holding steady at ~80% manning. I've never once seen a 12-man ODA outside of a CIF/CRF/CTAC. So 18X makes sense to build force structure directly. You have to be a little older to qualify for 18X because one of the hallmarks of SF is a more mature (and thus capable) force as we've discussed. In 2000, the average age on an ODA was 34 years with 12 years time in service. Those numbers today are 27 and 8. When we revived the 18X program amidst the GWOT surge it was originally intended to fill about 30% of the force. The thinking was that the force could absorb 30% and still maintain the maturity (read capability) level. But the actual number now is 50-60%. That presents some problems.

Because we value and rely on competence so much, it hurts when it's not there. An ODA is a little self-sustaining unit, and it requires everybody to pull their own weight and then some. Something as simple as running flat range training requires multiple moving pieces. You have to forecast, request, and draw ammo, the range itself, vehicles, medical coverage, training and risk assessments, weapons, etc. It's a complex little exercise with all sorts of associated tasks, and a flat range is about as simple a thing as an ODA can do. I had a Team Sergeant explain to me once how he had an 18X heavy team and he was prepping for a range day. He tasked a new guy to go get a vehicle and the guy had no idea how to dispatch, let alone PMCS, a HMMWV. So that task turned into a training event where an experienced guy had to stop what he was doing and train the new guy how to do that thing. Sand in the gears. Take that minor thing and multiply that by 1000 tasks and then do it in combat conditions in an austere non-permissive environment and you see where this might be a challenge. Don't forget that you're SF, but you're still in the Army and if you want to see some pucker factor then go ahead and screw up a sensitive item inventory, or lose track of the controlled meds, or have a guy roll through a vehicle inspection at the gate with some pyro that wasn't accounted for. Big Boy consequences.

Now, it's not the 18Xs fault that he doesn't have these experiences. He's just following the program. But even a soft-skilled E-4 from a low-speed unit knows how to dispatch a vehicle, how to do a layout, and how a key box works. Get too many guys who don't know enough, and you start to accept risk where you shouldn't. Risk to force and risk to mission. So, what to do? Isn't there a way to get guys schooled up? Can't you run some sort of indoctrination where you teach new guys these skills? Maybe. But that's one more thing that you've got to plan, resource, and execute that you don't have time or resources for. And what tasks do you not train so you can do this new stuff? Resources aren't infinite. Can you reduce the number of Xs to say 40%? Sure, but how many ODAs do you want to ghost? What missions do you want to decline? What about just tell Team Sergeants to 'make it happen!' Of course, all high-performing organizations employ the 'fuck it' model of management, right? There's no perfect answer.

And there's still this little thing called SFAS that tends to stifle lots of 'good ideas.' The selection rates are roughly 50% for Officers and NG, 40% for 18X, and 25% for ADE (Active Duty Enlisted) for an overall rate of ~36%. So you'd be inclined to think that your best bet to get selected is to go X. As an X you get the benefit of a guaranteed slot, you get SFPC, and you get to do it all with a little cohort of your buddies. Strength in numbers, right?

But the reality is a bit more nuanced. Officers select higher because they are, generally speaking, better prepared. You can't commission until 21 and you need at least 3 years before you can go to SFAS so your bare minimum age is 24, likely older. Officer culture, especially among junior Officers, demands physical fitness. Every commissioning source requires extensive land nav training. In other words, Officers already go through a pretty extensive prep. National Guard guys have the benefit of SFRE and likely some sort of unit sponsorship or even individual mentorship. That level of accountability and preparation creates better results.

Active-duty guys have to contend with all of the challenges of dealing with day-to-day unit life with all of the distractions so it's no surprise that they have the lowest success rates, right? Except the 18X numbers are cooked (the numbers are cooked, but nobody is cooking them - nothing nefarious here). There's no real way to track the actual success rates as the Xs are dependent on and inculcated into other systems that distort the real numbers, but a recent cohort broke down as such: 50 guys started OSUT. About 20 of those guys didn't keep their X contract through OSUT because of injury, PT failure, etc. So, 30 guys go on to Airborne School, but 5 don't make it. Again, injury, PT failure, etc. So now 25 go on to SFPC but another 5 drop out; fail land nav, fail PT, get injured, quit, etc. So, 20 go to SFAS. Of those 20, 8 get selected. So, the number looks like 40%; 8 of 20. Not bad. But in reality, it's 8 of 50 -- only 16%. Worse than Active-duty Enlisted. And you had every possible advantage. Now you're needs of the Army. Likely you'll end up as an 11B, likely in an Airborne unit. But no guarantee on your guaranteed contract.

For the record, every 18X (with a few years of team time) that I've ever met was just as switched on as a traditional Green Beret. No difference. This isn't a referendum on 18Xs, just planning considerations.

Officer vs Enlisted

I also get a lot of questions about enlisting or becoming an Officer. This again is a highly nuanced question, but you might find this discussion helpful. The fastest way to become a Green Beret is 18X. You'll get a contract that 'guarantees' your slot, loads of specific prep, and an instant community of like-minded guys. But it's not without its issues as described above. Lots of 18Xs have degrees; it's upwards of 50% have Bachelors. But it's not really a direct connection to the degree. A lot of Green Berets don't have a degree and aren't really interested in getting one. I'm not aware of any requirement to have one to get promoted and if there is a requirement, I'm not certain that's a good idea. Higher Education is a cesspool of effete liberals and narcissistic self-indulgent weirdos. It's bad enough that we make all Officers do it. You can be perfectly successful and spectacularly competent without a degree, just as you can with one. So, this isn't really a super relevant factor in your Officer vs Enlisted decision, it's really more about your operational timeline.

As an SF Officer you can expect to get 24 months of team time. You could extend that by getting a second team, but that's not really team time as those second teams are manned and employed differently for different missions. You could also go on to serve in a SMU, but that's an incredibly small percentage (SMUs typically select at a 5-10% rate). But there is a whole life beyond team time to consider. Everyone always defaults to 'staff time,' but not all staff jobs are created equal. Your worst day on staff in an SF unit is better than your best day on staff in a conventional unit. Plus, there are all sorts of jobs that are 'branch immaterial' or 'combat arms immaterial' that you can access and lots of them are incredibly rewarding. As an SF guy you'll be immediately competitive and uniquely qualified for them. Everything from intelligence, to operations, to advising, to teaching, to embassy work. You name it, it's out there. These usually aren't jobs available to NCOs.

Speaking from personal experience I stayed deployed and active doing SF missions for years after I left a team. I was exposed to all sorts of assignments from working with industry, to chasing drug runners around Central America with my own helicopter task force and host nation SWAT team, to planning $150M SOF shoot houses and range facilities, to planning multimillion dollar international exercises. I personally know guys that took their families on amazing embassy assignments to awesome posts, pursued advanced degrees, taught at Ivy League schools, developed weapons and gear, and many others. Missions that make team time seem quaint and boring to some.

There is the Warrant Officer option which can get you an additional 6-8 years of team time, but you can only access this once you get to SF, so it's not a primary planning consideration at this point. And I'd be doing a disservice if I didn't talk about the financials. As an SF NCO you do get some additional allowances, but you're still making significantly less than your Officer counterparts. In fact, over a normal 20-year career and subsequent retirement you can expect to make half as much. For most of us, it's not about the money, but the difference is significant.

I have surprisingly had more than a few guys talk about pressure from friends or family to go Officer because it's more "honorable." Let me just put that to rest definitively. Service is service and there is no difference in honor between Officer and Enlisted. That's some outdated and misinformed narrative of a time long gone. You may associate some level of prestige as an Officer, but you better put that attitude on a shelf quick. You serve at the convenience of your men. As an Officer, you are nothing without the men. If you walk into an SF unit with that attitude then the first guys to kick your ass will be the other Officers, while the NCOs get hydrated and warmed up.

As an Officer, during your team time you'll have all of the opportunities to train and attend the same schools that your guys do. You have the additional burden of training management and administration to distract you, but you are 'in the stack' just the same. In my experience, the Officers are expected to represent the full suite of SF skills to an even higher degree. How can you expect to lead men that can out-shoot, out-fight, out-ruck, out-run, and out-smart you? You've only got to keep up the pace for 2 years Captain, get after it.

But you don't necessarily get to be one of the boys. You're the Detachment Commander. You Command. You'll blow the froth off of many a cold one and have your back against the wall alongside them often enough. But you have the burden of leadership. You might have to make life and death decisions about these guys. You can't forget that. You can't outsource that responsibility. So, the decision between Officer or Enlisted is entirely personal. You need to decide what your best pathway is. There is no right or wrong answer and only you can decide.

Reading Lists and Journaling

Let's talk about the Reading Lists. I'm not sure what it is about Reading Lists, but this seems to be a topic of endless discussion and query. There are so many reading lists already published that it seems redundant to publish my own. The USSOCOM Commander maintains a list. So does the USASOC Commander. Even the JSOC commander publishes a list. They do this almost every year and you can easily find it with a little internet skill. Just search "USASOC Commanders reading list 2023" and off you go down the rabbit hole. It seems like most of the prominent SOF social media types have a list too. The crossover of the lists is exhaustive. There are enough lists to keep you in letters for years. Stop asking and start reading. I consider this horse dead. Well beaten. I'm much more interested in you doing some writing.

Journaling. That's right, I said writing. If you're an avid reader, you probably have a metric ton of ideas constantly flowing through your head. You want to make sense of these ideas, so you read more stuff. You get more ideas you struggle to make sense of. You read more. Rinse and repeat. You would be better served doing some writing. Writing forces you to organize your thinking. It makes you put your thoughts on paper, where those thoughts must be reckoned with. You can actually see what you think. You can debate the logic and the evidence and the merit. You can be judged. That's right, you can be judged. I think this is one of the inherent reasons most folks don't write. They're afraid of being judged. SFAS is a three-week long comprehensive judging process. You would be well-served to be comfortable being judged.

If you sat down and penned some thoughts about, say, "A Comprehensive Guide to SFAS," then people might judge you. If you got it wrong, they might laugh at you. If you were inaccurate, you might get corrected. If you were over-truthful, you might get exposed. Imagine putting that in a book and allowing anonymous people to judge you silently. But you should write your thoughts down. Make sense of what you read and put it to good use. How are you going to apply it? What is the next step? What could it mean if you combine it with this other thing you read? Check your notes? Cross reference. Make sense. Write it down. You don't have to share it with anyone, but if you aren't writing then your thoughts probably aren't making much sense.

I'll add to this the need for performance journaling. You can ease into this unfamiliar world with just keeping track of your workouts. That's manly, right? I can't tell you how many guys ask for advice but can't provide any real data to get started. If you want to get better, I need to know where you are or we're going to stay lost. Keep track of your workouts, then add in your nutrition data. How much water you drank. How much sleep you got. How do you feel? Not your feelings, but how do you feel? You did X workout with these times/weights/reps and you ate this and drank that and slept this. Now how do you feel? Are you getting faster, stronger, better? What if you adjusted this food or slept more or whatever? If you're not writing it down, then how are you keeping track of all this data? Are you serious about your performance or not?

If you keep a training log of your diet and exercise, add in some thoughts about some stuff you read, and add a to-do list, then you'd be performance journaling. Just like that you could organize your thinking. You could plan your growth. You could maximize your gains. You would get smarter and more formidable. If you are serious about building elite performance and you are not performance journaling, then you are not really serious about building elite performance.

Performance Enhancing Drugs and Nutrition

I am going to take a bit of a contradictory stance on this topic. I both support and discourage PEDs. I believe that we should have a military funded PED program for select units and individuals. I also think that taking PEDs as part of your SFAS prep is ill-advised. This is a difficult position to reconcile, especially given the increasing tolerance of PEDs, legal and illegal, in society today. But I think it is an important discussion to have. The nutrition argument is less obfuscated.

SOF has employed, to great success, PEDs and advanced and experimental treatments to help injured operators recover from injury on the battlefield. So, we have seen the potential benefits (Alford & Chang, 2020). We have also spent considerable cognitive horsepower studying this issue (Wigger & Oelschlager, 2017). There is an obvious moral argument to be made as well (Mehlmen, 2019). But I am looking at this from a performance perspective and a little bit of a business case. We want the best operators, with the most experience, with increased performance and longevity. If I have to turn to limited PEDs to get there, then so be it. As long as its medically administered, centrally resourced, and completely voluntary then I'm all for it.

When it comes to SFAS prep I'm a little more restrictive. It seems that supplements and PEDs are becoming an easy button or a shortcut to performance. I don't think that they represent the true self and I want to be able to assess the most accurate version of a Candidate as possible. And we must recognize that sometimes Candidates need to be protected from themselves. Men will die for points and getting selected is about a million points in the game of life. Many Candidates will be tempted to turn to PEDs in order to gain an advantage. But because we haven't culturally accepted this in the military, those Candidates will do this absent of proper oversight. Abuse, injury, and potential death are the likely outcomes. We have already seen this phenomenon at BUD/S training. That we are not openly addressing this aggressively in SF is near criminal negligence.

But a little pre-workout is fine and Candidates that are forced to eat at the Dining Facility will likely struggle with getting adequate nutrition. In these cases, then supplementing your diet with protein and creatine, at the correct serving sizes, is not just acceptable, its preferable. The standard American diet is atrocious. Seed oils, synthetic food, and processed crap is now the hallmark nourishment for most. The more that you can avoid these poisons and rely on clean whole foods the better your performance will be. If you are properly using your performance journal you will quickly correlate bad diet with bad performance and good diet with better performance. If you are looking for an advantage, start eating healthy and ease up on the booze.

Finally, I would highlight the importance of good rest. Both good rest as a deliberately programmed component of the fitness package and as a direct action in your sleep. Good sleep hygiene is critical to performance. Your muscle gains are started in the gym, but they are finished in the bedroom. Proper restorative sleep is perhaps the single most important factor in health and holistic fitness. Develop a proper sleep ritual by restricting screen time and avoiding stimulants just prior to going to bed. Buy a good pillow and mattress (or mattress topper for the barracks bound) and ensure that you restrict obtrusive noise and light whenever possible. If you can control the temperature, then cooler is better. Your body needs to drop a few degrees in order to enter restorative sleep (Harding, Franks, & Wisden, 2019).

So, some supplementation is acceptable. But before you start sticking yourself with needles or spending hundreds of dollars for a dubiously researched and heavily marketed powder, pill, or gel you might do a little work on the big stuff. Eat right, cut out the booze, focus on active recovery, and get good sleep. At selection you will not have any supplementation so if you can prepare without it, your days at Camp Mackall will be less of a shock to your system. Simple is always better.

No Shit, There I Was...Pissing Hot

Illicit drug use is obviously frowned upon in Special Forces. We reflect society so of course this vice makes its way into the Regiment, but it is definitely not the norm. Except this one time. My ODA was forward deployed to a joint host-nation military and police base on the edge of the Amazonian jungle. The base was along a key route from the Andean Ridge and was a hotbed for drug trafficking. The base would account for thousands of pounds of marijuana, cocaine, and precursor chemical seizures every month. Their procedure was to stockpile these seizures on the base and conduct a controlled burn once a month. A massive pile of drugs as big as a bus. Impressive.

One day we were sitting around the team house recovering after a grinding day of training and we noticed that along with our very relaxed posture there seemed to be a haze in the air. We went outside to investigate and discovered that the stockpile, which was just a few hundred meters from our house, was ablaze. The thick smoke was carried by the perfect wind right across our home. We were engulfed in the plume. We had unintentionally been hotboxing ourselves. Every single one of us was high as a kite.

On its face, this was bad. But we also had an administrative requirement to conduct a 100% urinalysis upon redeployment back to stateside. This was normally a perfunctory process, but now we were all at risk. I made a quick phone call back to Fort Bragg to talk to the Group lawyer. He advised us to document the event with lots of pictures, write up some official memorandum, and coordinate with the host nation forces to avoid another exposure. So, we did just that. We snapped a few dozen photos, got official memos from the host nation and even our co-located US DEA partners, and made certain to communicate our desire for notification of future events. Task complete.

Until the next month. We had a repeat of the exact same event. You might think that this is an indictment of our administrative acumen, but anyone who has deployed in these environments understands that our partner forces understanding of unit coordination differs significantly with US standards. This adaptability and flexibility are a hallmark of Special Forces missions. So, we repeated our documentation procedures and reinforced our desires with the host nation. Apologies were offered and plans were made. Task complete. Again.

Until the next month. 3 months in a row. Hotboxing that would make Cheech and Chong envious. I can joke about this now, but at the time this was a genuine crisis. A positive urinalysis test was presumptively guilty, and you could end a career over a single positive urinalysis. And we were all hot. We repeated our documentation procedures and finished out our deployment without further incident. We returned to Fort Bragg and went through our post deployment activities. My first task was to march myself up to the lawyer's office and submit all of our documentation. A few weeks later the urinalysis results were in. We had a 100% positive test result. But we also had 100% documentation. Bullet dodged.

All the rules matter. Until they don't.

Life Lesson: Own your mistakes and stay ahead of them. Taking ownership over your blunders is always better than trying to hide them. Time reveals all and you can't run away from your wrong doings, or they'll bite you in the ass.

Chapter 10: Cadre

I want to state this in the plainest language possible; nothing in my thinking or writing is a pejorative criticism of the SFAS Cadre. I have immense respect for what they do and how they do it. Their task is a nearly impossible one. If SFAS selection rates rise, then they get blamed for lowering standards. If SFAS selection rates drop, then they get blamed for gatekeeping. They serve many masters, and they have an incredible burden in that service. I have often related that duty as SWCS Cadre in general, but SFAS Cadre in particular, is perhaps the most important job in the Special Forces Regiment. A good Team Sergeant in the operational force will have a direct impact on ten or twenty guys. But a SWCS Cadre will have a direct impact on an entire generation of Green Berets during his tenure. SFAS Cadre are the first, and for ~64% of Candidates the only, exposure to Special Forces. They set the tone for how a Candidate perceives the Regiment. I still remember First Sergeant Callahan and Staff Sergeant Lowery. That is impactful.

I have spent innumerable hours and immeasurable miles with Cadre at SFAS. I have come to know them as some of the smartest, kindest, and honorable men in the

Regiment. They have impressed me as astute observers and perceptive assessors. They can elucidate on incredibly nuanced observations of Candidate behavior and performance and their ability to remain remote from the agony of the physical rigors is remarkable. They have been particularly impressive in their ability to remain impartial. During my directed observations of SFAS, I conducted an experiment on several occasions where I tried to get Cadre to make predictions of Candidate potential selection chances during the initial equipment layout. Never once would a Cadre even entertain the idea. Even as a joke. To a man, each Cadre would say that they didn't want to bias any future interactions with the Candidate. They were clear that they wanted to assess each Candidate on each event and only on the merits of their individual performance. I would describe their devotion to remaining impartial as almost monastic. They would neither encourage nor discourage Candidates.

But for Candidates, the Cadre remain a distant paradox. Candidates naturally want to impress, and they look for feedback to gauge their efforts. Cadre remain detached and repeatedly answer questions with the familiar, "Do what you think is best, Candidate." Rarely do Cadre smile and Candidates start to think that Cadre are indifferent, that they don't care. I have seen how much Cadre care in the way they react to an injured Candidate. Occasionally, Candidates will drop an apparatus. It usually falls harmlessly to the ground. But there is little that will generate a response from Cadre quicker than a dropped apparatus. Cadre rarely yell, but a dropped apparatus will almost certainly earn you immediate, sharp, and loud words. The apparatus can be dangerous. Upwards of a thousand pounds crashing to the ground will snap bones with painful ease. I've seen it happen.

    During an unassuming low carry event a weaker Candidate dropped the apparatus, and it caught the calf of a teammate. The stricken Candidate was a clearly strong performer and was a lock to be selected. I will never forget the look of complete desperation on the face of the injured Candidate. He was laying on the ground, leg pinned and clearly broken, and reaching up to the Cadre. It was a look of terror, agony, and realization as he was immediately coming to terms with the fact that he was done. Selection was over. The Cadre jumped into action, pushed through the loitering Candidates, and single-handedly pulled the apparatus up with one arm and dragged the Candidate out from underneath with the other. A feat of raw strength and paternal compassion. I could see the rage in his eyes. He calmly called for a medic, comforted the traumatized Candidate, and gave direct commands to the other Candidates to continue training.

As soon as the medics arrived, he pulled back away from the Candidates, and we shared a moment. He was shaken. Not from the event itself, rather he was shaken because he knew that the now-dropped Candidate was almost Selected. He felt somehow responsible. It was clearly not his fault. But he cared. Candidates think cadre don't care. Nothing could be further from the truth. Cadre do care and they are deeply passionate. Not passionate about being accessible or likeable. They are obsessive about standards. They know that the very future of the Regiment is in their hands. They have made a sacred pact with the Regiment, and they are zealous in that commitment.

I have touched on this subject earlier, but I want to call specific attention to race, color, religion, national origin, or sex. They simply don't matter to Cadre or to the Regiment. In an age when diversity, equity, and inclusion are used as a cudgel for ideological evangelists, these categories are just part of the landscape at SFAS. The lack of diversity in SOF has been a topic of interest for decades. Early research cited institutional barriers like ASVAB score cutoffs and swimming requirements (Harrell, et al., 1999). Do we really want to encourage less intelligent and less capable operators? Recent studies have been much more pragmatic noting that diversity for diversity's sake is unwise. You can encourage diversity at the enterprise level, but SOF combat formations cannot be *made* to be more diverse (Simons, 2019). In my more than two decades in Special Forces I have never once seen a single instance of racism. If you wear the wrong tactical khakis you will most certainly hear about it, but race is simply not considered.

None of this matters at Selection. Only your performance matters. If you are a minority or you feel marginalized, know that you will be given the exact same treatment and consideration at SFAS that every other Candidate gets. No favoritism and no bigotry. The deck is not stacked against you. There is no deck; just you and your performance. The same can be said for your Military Occupational Specialty (MOS) or special schools or college degrees or any of the dozens of other qualifications you may have. None of it matters to the Cadre. You don't have to worry about representation or perceptual barriers such as lack of knowledge or community support. SFAS is the ultimate meritocracy.

This recalls my earlier description of Cadre as monastic. They have taken a sullen vow to be impartial assessors. Cadre go out of their way to know as little of your background as possible. As a senior researcher, I was often provided a roster of Candidates to include rank and MOS to aid in my observations. Whenever I referenced that roster while walking with Cadre, they would almost recoil from it. They remained steadfast in their commitment to impartiality. It really is remarkable considering the attention this topic seems to get in the real world and especially online, which is most definitely not the real world. Camp Mackall is its own little upside-down universe. In fact, the mantra of neither encourage nor discourage would sometimes present its own issues of neutrality.

A phenomenon that we observed repeatedly was interactions with female Candidates. Following the lifting of the Combat Exclusion Policy in 2013, female Candidates have been allowed to attend SFAS. As one might imagine this was a significant cultural shift for Green Berets. To their great credit, SWCS has maintained the fair, unbiased, and impartial training environment that Selection has always represented. The Cadre have been the most critical component in this, and they deserve to be lauded for their efforts. Other male Candidates on the other hand were less impartial. We repeatedly observed male Candidates give preferential treatment to female Candidates.

During multiple observations under multiple conditions, I observed male Candidates demonstrate preferential treatment to female Candidates. For example, male Candidates were less likely to verbally call out weak physical performance from female Candidates than from male Candidates. Likewise, male Candidates were more likely to physically assist weak physical performance from female Candidates than they were to physically assist a male Candidate's weak physical performance. These observations are most obvious during Team Week. Occasionally, teams who underperform are assigned additional 'corrective training.' A team that misses a time hack might have to perform extra push-ups or the like. One particular event directs that Candidates hold their full ruck over their head, in an arms-locked position, for 30 seconds. The time starts when the entire team is in the correct position and is terminated only if every member of the team maintains that position for the full 30 seconds.

Female Candidates, on average, exhibit less upper body strength than male Candidates. As such, many female Candidates would struggle with this corrective training.

Interestingly, when a team had both male and female Candidates struggle, the other strong male Candidates would literally rally around the weaker female Candidates and physically assist them with their ruck. At the same time, the team would actually turn their backs on a struggling male Candidate. I observed this phenomenon repeatedly and consistently, which led me to consider Candidates' socialization as the biasing factor to assessment and selection. When I interviewed Candidates about this behavior, male Candidates were often unaware of their actions, noting that they were just "acting naturally." Again, none of this matters to Cadre. They remain impartial observers. Only your performance matters.

As a Candidate, the Cadre are not your friends. Remember, Camp Mackall is an upside-down world. These Green Berets are not there to be your drinking buddy or to teach, coach, and mentor. They are not traditional NCOs in your chain of concern. They are there to assess and select. You would be well-served to remember this. When you get to an ODA you will find no more loyal and personable fella. At SFAS, every single interaction that you have with the Cadre is an opportunity to observe your performance across the physical, cognitive, and interpersonal domains.

Do not ever let your guard down and do not expect to be treated as anything but a Candidate. You are there to perform and they are there to assess. You will be treated fairly, but you will be treated with indifference. They invited you to come to SFAS, but you also volunteered. I guarantee that the Cadre will not forget their role. Neither should you. PC or Beret.

No Shit, There I Was...The Punk Bunk

This is a story from Ranger School, but it details a telling interaction between Green Berets, and it is a poignant lesson that is worthy of emphasis. We were on one of our final patrols in the Florida phase. At this point in the course the Ranger Instructors (RIs) are much more relaxed. There is a recognition that students are both physically and mentally exhausted and they are also much more proficient, so they don't require as much external motivation. Students still need supervision, but there is much more leeway given.

My buddy Jeff was the Weapons Squad Leader, and I was just a lowly ammo bearer on one of his gun teams. But because we were Green Berets, we had a sacred duty to each other. One night as we made our security halt before establishing our patrol base, aka the Punk Bunk, I violated that sacred duty. I fell asleep. Falling asleep in Ranger School is par for the course. I'll admit that I'm a little more 'solar powered' than most, but everyone dozes off from time to time. That's my defense and I'm sticking to it. I recently learned that the term Punk Bunk is no longer common vernacular. I think this is an erasure of history and I hereby submit this manuscript in an attempt to reintroduce the term Punk Bunk in lieu of Patrol Base. Let it be so.

At any rate, all I remember is waking up with Jeff standing over me, beating me up, and screaming at me. And the sun was up. Those that understand the little patrol base dance understand the timing and priorities of work that go into these things. Suffice it to say, the sun being up and you sleeping outside of the punk bunk is a bad thing. Apparently, my gun team had fallen asleep at a critical juncture the previous evening and had not moved the final 300 meters to our patrol base with the rest of the platoon. The platoon was deep into conducting punk bunk activities, and Jeff had spent that past several hours looking for us. Losing a gun team is a big deal. The kind of big deal that can get you recycled, and we were just days away from graduating.

Jeff was rightfully pissed, and he was *not* struggling to communicate this to us. We grabbed our gear and hauled ass up to the patrol base where we found our platoon just finishing the morning stand-to alert procedures. Jeff told us where to go and ran off to attend to his other duties. We started feverishly digging in a hasty fighting position.

Somehow, the Ranger Instructors had not noticed our absence and we were eager to keep it that way. As we were scraping out a little hole in the Florida sand an RI walked up.

"Hey Ranger! What the fuck are you doing?"

I responded, "Sarn't, our squad leader told us our position wasn't up to standard so we're improving it since stand-to is over." Technically correct.

The RI yells out, "Weapons Squad Leader, get over here!"

Jeff bounds across the patrol base, and I can see the vitriol in his eyes. "Yes Sarn't"

"Did you make these Rangers re-dig this position?"

"Sarn't, once the sun came up, I noticed that their position wasn't up to standard." Technically correct, again.

The RI pauses, then bellows out, "Out-fucking-standing!" then he yells even louder so that the whole platoon can hear him. "Hey Rangers, this is the fucking standard! Too many of you think Ranger School is already over. These Rangers are still in the fight. Ya'll better learn something from these SF guys!"

He then issues us all *major plus* observation reports. We went from complete dirt bags who fell asleep and endangered Jeff's very existence to rock stars with paperwork to prove it. Zero to hero. We sign our reports and the RI walks away. Jeff turns to us and says, "Thanks, but if you ever do that again I'll fucking kill you." Punk Bunk justice.

Always look cool.

Life Lesson: Never stop working. Even when things look desperate you can always improve your position. You can often snatch victory from the jaws of defeat with a little persistence. And sometimes everyone needs a little luck.

Chapter 11: The Standard is the Standard

There is always a prevailing discussion of standards surrounding SFAS. The general scheme is that whatever SFAS is now, it is somehow easier than it once was. Everyone seems to have attended the last hard class. This discussion of eroding standards is worth spending some time debating. This harkens back to our earlier discussion about shaping the prevailing narrative. The moment something at SFAS changes, it is immediately perceived to be a lowering of standards. This message is repeated, without evidence, until the narrative becomes the truth. SWCS is in a very difficult position in this regard. The current best practice is to say nothing, to neither confirm nor deny. There is an institutional fear of saying too much, revealing too many methods and standards, so they simply say nothing. The real shame in this practice is that it sacrifices the very people that it is meant to protect, the Cadre. The SWCS Cadre, particularly the SFAS Cadre are in an unwinnable situation. As previously discussed, if selection rates go up, it must mean that the Cadre have lowered the standards and are just giving it away. If selection rates go down, it must mean that the Cadre are gatekeeping and unfairly penalizing Candidates for minor infractions. Damned if you do and damned if you don't, and in this case neither situation is remotely true.

Since SWCS can't tell the story, I will. The standards at SFAS have not been lowered. I have been studying SFAS, from the inside, for nearly two decades and I have seen all the evidence from its inception in 1988 until now. The standards have not been lowered. SWCS is in a bit of a conundrum, however. They can't really prove this assertion unless they release SFAS data. You can't protect the standards, the very integrity of SFAS, without protecting this data. Conversely, you can't make the argument that the standards have been lowered without the data either. So, we go back to narrative. This might be a good point to talk about the Taxonomy of Information. This scale of reliability is a simple way to classify information to help us understand complex problems.

The Taxonomy of Information			
	True	Relevant	Exclusive
Narrative	X	X	X
Facts	✓	X	X
Data	✓	✓	X
Evidence	✓	✓	✓

Narrative seems to dominate our discourse. But narrative is often not true, and when it is true it is often closer to war story true than fully true. It's a helpful tool to steer sentiment, but it is often not based in fact. Facts are true, but they are often irrelevant to the point being discussed. This is frequently the realm of the logical fallacy of Appeal to Authority. Just because you have an advanced degree, or you work in a certain organization may be true, but they are points often irrelevant to the discussion at hand. Data gets a bit more specific wherein it is both true and relevant, but often not exclusive; there could be other explanations. Evidence is the goal of these discussions. It is true, it is relevant, and it only supports the specific argument at hand. Argumentation is both parts logic and evidence, or as close to evidence as you can reasonably get.

But even knowing this, SWCS is still stuck with the unenviable position of fighting the upper left of the chart dynamics and they simply cannot use the lower right tools. Since we cannot release the data, the real evidence, let's use logic to shape the story a bit. Let's fight fire with fire with two of the most cited tales in the narrative of eroding standards, the gender-neutral PT test and Log and Rifle PT. These two events are an interesting case study into how good decisions take on a life of their own and get distorted because SWCS doesn't follow up with a deliberate public information campaign.

In 2016, when SWCS announced that the Physical Fitness Assessment in SFAS would become gender neutral, there was a cry from the uninformed that this was clear evidence that the standards were being lowered. We should examine this closely. First, the SFAS Physical Fitness Assessment (PFA) is not the Army Physical Fitness Test (APFT). Both events often get called the "PT Test" which further muddies the waters. Do you recall our discussion earlier about being clear in our language because words have meaning, and those meanings can be impactful? This event is unique in SFAS in that it is the only event where Candidates know the standard. It is well advertised, and Candidates are required to report to SFAS with a passing test score given within 30 days of reporting. The published standard was a minimum of 49 push-ups, 59 sit-ups, and a 2-mile run time of 15:12.

The PFA also includes six pull-ups. The APFT has none. I would suggest that the inclusion of the pull-ups would prove a massive barrier to entry for female Candidates and a general indication of increased standards, not lower. I cite the Leg Tuck data from the 2022 Rand *Independent Review of the Army Combat Fitness Test*. This report revealed that alarming numbers of female Soldiers were unable to complete even one single leg tuck. Is a leg tuck a pull up? No, but any reasonable person can surmise that the leg tuck is the easier starting movement for a pull-up and if you cannot do even one leg tuck you are unlikely to complete one pull-up, let alone six. I understand that the leg tuck stresses the pelvic floor in ways that the pull-up does not, but we are just following some simple logic here. If you are not willing to make this obvious connection, then you are not willing to engage in good faith argument.

The Rand study showed that 28% of active-duty female Officers and 48% of active-duty Enlisted women could not complete the leg tuck requirement. Within the Guard and Reserve population nearly 60% of enlisted Reserve and Guard women failed the ACFT and 51% of female Reserve Officers and 43% of female Guard Officers failed (Rand, 2022). From this data, we can logically conclude that the simple addition of pull-ups into the PFA is clear evidence that not only have the standards not been lowered, but the standards have actually gotten harder. That's narrative crushing evidence.

We should continue our examination of the gender-neutral PFA standards with the first event of the PFA, the push-ups. Under a gender segregated APFT, the minimum score for female candidates would have been 25 push-ups. Under the gender-neutral construct at SFAS the minimum score is the same as male candidates, 49 push-ups. So again, the gender-neutral standard is harder for female Candidates in relation to the gender segregated standard. This is such an obvious conclusion that it begs the question of just how informed the authors of the falling standards narrative are. The evidence thus far is clear, the standards have gotten harder – not easier. I should note that in the Fall of 2022, the standard-pushup has been replaced with the hand-release push-up and the minimum standard is 29 repetitions. For both male and female candidates. One standard.

The next event is sit-ups. Sit-ups are unique in that there was no difference in male and female APFT scores. The SFAS PFA standard was 59 repetitions. I say *was* because in the Fall of 2022 the sit-ups were dropped from the PFA. SWCS cited the growing body of evidence that sit-ups measure hip flexor strength primarily and were not an accurate measure of core strength and the exercise represented a significant risk of injury. Interestingly, sit-ups were the one event in the PFA that was most often failed. Failed by male candidates. I can provide no evidence of this as this data is not available for public release. I would ask SWCS, what harm is there in releasing PFA data? Since sit-ups were removed, we haven't heard the outcry about tumbling standards that we did when the gender-neutral standards were announced. This again might make us question the credibility of the authors of the falling standards narrative. There are no formal plans to replace the sit-ups with another event.

The third PFA event is the 2-mile run. This standard is set at 15:12 and again destroys the gender-neutral argument because the gender-segregated standard allowed a full 18 minutes for female Candidates to reach the minimum score. Yet again, transitioning to the gender-neutral standard effectively raised the standard for female Candidates. It decisively did not lower the standard for male Candidates. This is glaringly clear evidence. How did this argument ever get any traction at all? And this discussion doesn't even consider that the PFA is just an admittance exam. You could remove the entire event from the training schedule and all you would do is get two dozen more students into the first week. They still have to complete the rest of the Selection.

Additionally, the studies show that high PFA scores correlate with high selection rates, and conversely low PFA scores correlate with low selection rates. This data is evident from some of our earliest available studies in 1990 where there was a clear linear relationship between APFT (PFA) scores and probability of success in SFAS (Teplitzky, 1990) and remains clear all the way through 1999 where the specific rates were recorded at a 57% selection rate at 254 and above and 43% rate below 254 (Zazanis, Hazlett, Kilcullen, & Sanders, 1999). This data trend continues to 2019 (Farina, et al., Physical performance, demographic, psychological, and physiological predictors of success in the U.S. Army Special Forces Assessment and Selection course, 2019). So the evidence is clear that you must score well on the APFT/PFA to be successful at SFAS, but removing the event or the score, eliminating that specific assessment, doesn't relieve you of the reality that high levels of fitness are required. How this empty argument persists simply defies logic.

Log and Rifle PT, inexplicably, also remains a hot-button topic among many Green Berets. The event embodied the toughness of SFAS until the late 2010s. Log and Rifle PT, conducted near the end of the first week of Selection, represented a departure from the normal training environment. SFAS Cadre follow a neither encourage nor discourage disposition towards Candidates remaining stoic in the face of the often-chaotic environment. SFAS Cadre simply issue indifferent instructions and then sit back and record the results. Contrast this with nearly every other school in the DoD where cadre take a much more adversarial stance. Log and Rifle PT in SFAS adhered to this traditional DoD model, but with the intensity one might expect of the SFAS setting. This three or four-hour event was a feeding frenzy. Cadre would engage in the traditional 'shark attack' mentality and move up and down the ranks offering biting criticism and poignant commentary on Candidate performance. Colorful language was the norm. Ironically, most Candidates reported that this was a bit of a relief as they viewed this event as simply a survival event rather than a performance event. Indeed, there were no documented performance objectives other than to simply survive. There is no recorded longitudinal data on the event, but I have interviewed multiple long-time SFAS staff and medical support personnel, and they report that Log and Rifle PT historically lost more Candidates to injury than to quitting. Even then the numbers were incredibly low, accounting for only two or three students per class. And other than perseverance, there was no measurable attribute to assess.

The Log and Rifle PT event was dropped. It was replaced with the Combat Readiness Assessment (CRA) which is a series of quantifiable performance tasks including a sled drag, rescue dummy lift and carry, and heavy ammo can movement. The CRA provided much more valuable assessment data and from an analytical standpoint is much more useful. But it doesn't have the 'suck factor' that Log and Rifle PT had, and it was change. Human nature is predisposed to dislike change and the Green Beret community is no different. The story became that because there was a change at SFAS it must mean that the standards are compromised. But how would removing an event with no performance standard other than *survive* represent a lowering of standards? What is the logical leap required to make that argument? I acknowledge that SFAS with Log and Rifle PT is *different* than SFAS without Log and Rifle PT, but is it harder? A reasonable argument can be made that the CRA is quantifiably harder than Log and Rifle PT.

Does this mean that we can make the argument that since the new CRA SFAS is quantifiably harder than the Log and Rifle PT SFAS and that all of those generations of Green Berets that assessed under the *easy* model didn't meet the standard? If that is the case, then we can follow this hypothetical across the history of every difference in SFAS. When I attended SFAS, pull-ups were not part of the APFT. They were randomly tested across one of many possible locations. In my class, they were administered immediately following the Nasty Nick obstacle course. And I mean immediately. You crossed the finish line dripping in mud and panting from exertion and were instantly instructed to mount the bar. Cadre stood directly below the bar and if you so much as grazed the Cadre the event was terminated. I don't recall any Candidate performing more than two repetitions. Is my Green Beret compromised? I found eight points during my Land Navigation test; can I conclude that anything less than eight is an erosion of standards? We could follow this illogical logic in perpetuity until we locate every single difference and confiscate every single beret. Are we beginning to see the folly of policy by narrative vs policy informed by data and evidence? We are tilting at windmills.

No Shit, There I Was...The Don't Quit Myth

There seems to be an enduring myth that all you need to do to pass SFAS is not quit. This is a gross misrepresentation of the *don't quit* mantra. Yes, nobody who quit was ever selected and everybody who got selected didn't quit, but this is not the same thing. Performance at SFAS is the only thing that matters. You must perform. You could have all the heart and resilience in the world, but if you can only crank out 18-minute miles under a ruck then you are not going to make it.

The logical conclusion to this *just don't quit* mentality is that many Candidates in training think that they will rise to the occasion. They believe that somehow, when the moment requires it of them, they will marshal their inner bad-ass and somehow make it happen. It doesn't matter what *it* is, they will just somehow do it. This often manifests in poor training habits. Practice does not make perfect. Practice makes permanent. If you engage in your SFAS prep in some haphazard manner with the false belief that you will rise to the occasion at Camp Mackall then you will become one of the 64% of Candidates that don't get selected.

There have been multiple studies that attempt to quantify this phenomenon. In 2010 the most comprehensive study was published, and it concluded that perseverance was so weakly correlated with success that it should not be measured on its own. Rather, it should be incorporated with other predictors (Beal, 2010). This same study also notes that while perseverance was correlated with successful selection, it was also measured highly in non-selects. Essentially, nobody admits to being a quitter. Quitters don't last at SFAS, but if your SFAS prep plan is to improvise your fitness program, hope for land navigation deliverance, and roll the dice on the Sandman, then you will get exactly what you prepared for.

If the *don't quit* mantra was so valuable, then why do we have 21-day non-selects? These unfortunate Candidates are the living embodiment of just *not quitting* not being enough. Remember Pig Pen? Please stop saying this and if you receive this advice from someone else, then you can safely ignore them. Focus on tangible preparation and leave the superlatives to others.
Performance is the only thing that matters.

The Last Chapter: The Why

When we started this book, I told you to think deeply about your *Why*. I hope that you have a *Why* now, or that you are starting to form one. But I need you to understand a few things going into this Special Forces endeavor. This is important, so pay attention. There are two things that you need to internalize. There are two things that are just general certainties that most guys seem to blindly discount and blatantly ignore. They are doing themselves a massive disservice when they disregard these things. These two universal truths are that you probably won't make it and if you do make it, you probably won't be doing the things that you think you will.

You are more likely to falter than prosper. That's the reality. You have a 36% chance of getting selected. That's it. No matter how we structure SFAS, no matter what events we adjust, and no matter when you go, it's just 36%. Some stretches we average in the mid-forties. Some stretches we average in the mid-twenties. But those upsurges and declines are always short-lived, and the historical average is 36%. Those are your odds.

I firmly believe that this book has everything that you need to increase your chance of success. It is all here in these pages. Stronger than the fastest runner, faster than the strongest lifter. A rock-solid rucker. Smart, capable, and adaptable. You are good at land nav, able to tie knots, and a skilled and effective communicator. You can give constructive criticism and you can handle being criticized. You can be affable, and you can be firm. We are going to judge you harshly. We have created the most demanding 21 days that you could possibly hope for, and we are not changing our standards. I want you to be in the 36%, but the reality is that you are far more likely to be in the 64%. You better have a plan for that reality.

Isn't that a loser's mentality to plan for failure? First, I wouldn't consider being a non-select failure. If you come to SFAS and lie, cheat, or steal then you failed. But if you came and just didn't make it, then you just didn't make it. Most don't. You need a fair bit of luck to get selected. If you get bad weather, it makes an impact. I did an informal analysis of rainy weather during land nav, and my data indicates a 10-20% increase in land nav drops when it rains, particularly in the winter. Injuries also happen. Even if you don't get a med drop, which you likely won't, you can still get injured to the point that it cascades into a Selection ending condition. You could plan a bad route on the STAR or get placed on a weak team for the Sandman. You could fall off the rope on the Nasty Nick. So, a little luck is helpful. But the odds are not in your favor.

Even with perfect weather or remaining injury free, SFAS is as hard as it gets. It's designed to be selective. We are going to put you in the worst possible conditions and ask you to perform to nearly impossible standards. And we won't even tell you what those standards are. A three-week long gut punch. With a ruck on your back. And an apparatus. And a disabled jeep. In sugar sand. Day after day after day. SFAS is hard.

So, you better have a plan if you don't make it. Plan number one should be to thank the Cadre, thank your fellow Candidates, and then shut your fucking mouth. Don't piss and moan. Don't complain. You knew what you were getting into. You volunteered and you knew the odds. Take responsibility for this situation. Plan number two should be to get your ass back to work and come back. Return to Camp Mackall and give it another shot. If your *Why* is righteous, then it should be intact. If you were truly committed to this endeavor, then you should still be committed. If it was worth the sacrifice for one shot, then it should be equally worth it for two. Or three. Come back.

Understand that no matter the reason for not making it, you are going to suffer a moral injury. It is going to suck watching your buddies move forward without you. You are going to blame everything and everyone. Including yourself. You will question your commitment, your intelligence, and your fitness. You will doubt your methods, you will blame this book, and you will struggle with this new reality. My advice to you is to internalize this. Train smarter. Get more focused. Work on your communication skills. Get your feet harder. Put in the work and come back to Camp Mackall. You have a duty. I like to think that the guys who are made for this life are already Green Berets in a way. They just have to come out to the Land of the Longleaf Pine and prove it. So come back and prove it. To us and to yourself.

When you do get selected, you need to understand that you won't be doing what you thought you would. Most of you think that Special Forces is all about kicking in doors and shooting bad guys in the face. Switching on your quad-tube night vision goggles, fast roping from a helicopter, tossing a flash bang, and clearing a room. Cool guy freefall stuff where you get to do a flip off the ramp of the aircraft and fly on down to the X. Ripping around the desert in a tactical dune buggy. Sexy gear and lots of Velcro. Instagram worthy.

You have been raised on images of the Global War On Terror and you follow all the wrong social media accounts. You have a perverted understanding of what Green Berets really do. You want that beret. But you need to wear that PC. The Global War On Terror is over and with it are the ubiquitous kill/capture missions that came with it. Even at the height of the Global War On Terror, Green Berets did ten key leader engagements for every one kill/capture mission. We gave far more immunization shots than kill shots. We can absolutely flip that switch and return to that mode of operation, but it is no longer the norm. Green Berets are a force of nature, but we are a better force multiplier.

So, your *Why* should include a good amount of teaching. You need to be a language and culture expert. Learning a foreign language is important enough, but you need to become an amateur anthropologist as well, because if all you can do is speak in the native language then you're more likely to insult with a perfect accent than you are to be truly culturally competent. You need to learn how to task organize a classroom and analyze a lesson plan for outcomes and assessments. You need to become a master planner. You must comprehend the nuances of mission analysis and you must grasp the details of branches, sequels, and contingencies. None of that stuff looks sexy on social media, no matter what filter you use. But you must master it nonetheless because that's what Green Berets do. Being elite means that you never say, "Can I?" You always say, "I can." And then you do it.

So, figure out your *Why* and then come out to Camp Mackall and prove that you found it. Eight times a year we host Special Forces Assessment and Selection. Eight times a year we invite the best that this Nation has to offer out to the Pinelands of North Carolina. We have built a little upside-down world where we will test you to the absolute limits of physical, cognitive, and interpersonal performance. We aren't lowering our standards. We are holding the line. We always have. Come join the Brotherhood.

Ruck up or shut up.

De Oppresso Liber

Bibliography

Alford, C. A., & Chang, S. (2020). *Net Benefit of Performance-Enhancing Drugs Within U.S. Army Special Forces.* Monterey: Naval Postgraduate School.

Bartone, P. T., Roland, R. R., Picano, J. J., & Williams, T. J. (2008). Psychological Hardiness Predicts Success in US Army Special Forces Candidates. *International Journal of Assessment and Selection*, 78-81.

Beal, S. A. (2010). *The Roles of Perseverance, Cognitive Ability, and Physical Fitness in U.S. Army Special Forces Assessment and Selection .* Alexandria: Army Research Institute.

Burke, A., Peirce , B., & Salas, C. (2006). Understanding adaptability: A prerequisite for effective performance within complex environments. *Advances in Human Performance and Cognitive Engineering Research.*

Defense, D. o. (2020). *2020 Qualified Military Available (QMA) Study.* Washington DC: Department of Defense.

Duffy, B. J. (1999). *Core ideology can be found in SF history.* Fort Bragg: Special Warfare Magazine.

Farina, E. K., Thompson, L. A., Knapik, J. J., Pasiakos, S. M., McClung, J. M., & Lieberman, H. R. (2019). Physical performance, demographic, psychological, and physiological predictors of success in the U.S. Army Special Forces Assessment and Selection course. *Physiology & Behavior*.

Farina, E. K., Thompson, L. A., Knapik, J. J., Pasiakos, S. M., McClung, J. P., & Lieberman, H. R. (2022). Anthropometrics and Body Composition Predict Physical Performance and Selection to Attend Special Forces Training in United States Army Soldiers. *Mil Med*, 1381-1388.

Feeley, S. (1998). *Special Forces Assessment and Selection.* Monterey : Naval Postgraduate School.

Hagerman, B. (1997). *Hagerman, Bart, ed. (1997). USA Airborne: 50th Anniversary. Turner Publishing Company.* Turner: Turner Publishing Company.

Harding, E. C., Franks, N. P., & Wisden, W. (2019). The Temperature Dependence of Sleep. *Frontiers in Neuroscience*.

Harrell, M. C., Kirby, S. N., McCombs, J. S., Graf, C. M., McKelvey, C., & Sollinger, J. M. (1999). *Barriers to*

Minority Participation in Special Operations Forces. Washington DC: Rand Corporation.

Knapik, J. J., Farina, E. K., Ramirez, C. B., Pasiako, S. M., McClung, J. P., & Lieberman, H. R. (2019). *Medical Encounters During the United States Army Special Forces Assessment and Selection Course.* Mil Med.

Lujan, F. (2013). *Wanted: Ph.Ds Who Can Win A Bar Fight.* Washington DC: Center For A New American Security.

Maladouangdock, J. (2014). *The Role of Strength and Power in High Intensity Military Relevant Tasks.* University of Connecticut.

Mehlmen, M. (2019). Bioethics of military performance enhancement. *BMJ Military Health*, 226-231.

Mission: Readiness. (2009). *Ready, Willing ,And Unable To Serve.* Washington DC: Mission: Readiness.

Mueller-Hanson, R., Wisecarver, M., Dorsey, D., Ferro , G., & Mendini, K. (2009). *Developing Adaptive Training in the Classroom.* Arlington: U.S. Army Research Institute for the Behavioral and Social Sciences.

Murphy, J. (2019, October 23). The Colorful and Controversial History of the Army's Berets. *Audacy*. New York, New York, USA: Connecting Vets.

Parrish, M. C. (2013, March 18). *Welcome to the 'Nasty Nick'*. Retrieved from Army.mil: https://www.army.mil/article/98879/Welcome_to_the_Nasty_Nick/

Rand. (2022). *Independent Review of the Army Combat Fitness Test: Summary of Key Findings an*. Washington DC: The Rand Corporation.

Richmond, K., & Roukema, L. (2010). *Identifying and Validating a Model for Special Forces Training Alignment & Performance Success*. Fayetteville: Tailored Training Products.

Scott, A. (2015). Ruck Deep Dive – Study #1: Physical Attributes Which Relate to Rucking. *MTI Research*.

Scott, A. (n.d.). Ruck Deep Dive: Study #2 - Ruck Training Adaptation.

Scott, A., Shaul, R., & McCue, S. (2015). Ruck Deep Dive: Study #2 - Ruck Training Adaptation. *MTI Research*, 1-11.

Simons, A. (2019). Diversity and SOF: Boon or Bane? *Special Operations Journal*, 42-52.

Teplitzky, M. L. (1990). *Physical Performance Predictors of Success in Special Forces Assessment and Selection.* Alexandria: Army Research Insitute.

United States Special Operations Command. (2018). *USSOCOM Implementation Plan Progress.* Tampa: United States Special Operations Command.

USAJFKSWCS. (2020, January). Special ForcesDetachment Mission Planning Guide. *GTA 31-01-003.* Fort Bragg, NC, USA: USAJFKSWCS.

USAJFKSWCS. (2021). *SFAS 02-22 Daily Summary.* Camp Mackall: USAJFKSWCS.

USAJFKSWCS. (2022). *SFAS 07-22 Daily Summary.* Camp Mackall: USAJFKSWCS.

USAJFKSWCS. (2022). *SFAS 08-22 Daily Summary.* Camp Mackall: USAJFKSWCS.

Velky, J. (1990). Special Forces Assessment and Selection. *Special Warfare*, 12-15.

Wigger, C. G., & Oelschlager, P. J. (2017). *The Moral Obligation to Explore the Military Use of Performance-Enhancing*

Supplements and Drugs. Monterey: Navl Postgraduate School.

Zazanis, M. M., Hazlett, G. A., Kilcullen, R. N., & Sanders, M. G. (1999). *Prescreening Methods for Special Forces Assessment and Selection.* Alexandria: Army Research Institute.